U0143221

Surveys in Geometric Analysis 2023

（几何分析综述 2023）

Editor-in-Chief: Gang Tian（田刚）

Edited by Qing Han（韩青）& Zhenlei Zhang（张振雷）

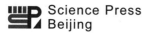
Science Press
Beijing

Gang Tian
Beijing International Center for Mathematical Research, Peking University,
Beijing, China

Qing Han
Department of Mathematics, University of Notre Dame,
Notre Dame, USA

Zhenlei Zhang
School of Mathematical Sciences, Capital Normal University,
Beijing, China

Copyright© 2024 by Science Press
Published by Science Press
16 Donghuangchenggen North Street
Beijing 100717, P. R. China

Printed in Beijing

ISBN 978-7-03-078965-5

Prologue

I am delighted that the *Surveys in Geometric Analysis 2023* will be published soon. Geometric analysis has been a very important and active research field at the frontier of mathematics for the last few decades, and many important achievements have been made in this field. The development of geometric analysis has played a key role in establishing numerous basic theories and solving long-standing problems in mathematics. It is a field worthy of much continued explorations.

Since the late 1990s, Peking University has taken the lead in organizing the annual conference in geometric analysis with participations of universities and research institutions both domestically and internationally. Scholars from all over the country and also abroad gathered together to report and share their own research works and exchange their ideas and thoughts. These have greatly promoted the development of geometric analysis research in China. The articles in this volume are exemplary submissions from the 2023 Geometric Analysis Conference, showcasing the progress made by Chinese scholars, both domestic and overseas. Readers will find in this book an opportunity to quickly grasp current frontier problems in geometric analysis, acquaint themselves with the latest research methods and results, and find valuable references for further research.

This book is the seventh volume of collected papers published since the birth of the "Geometric Analysis Conference" nearly 26 years ago. We extend our sincere gratitude to Science Press for their support in the publication of these collected papers. We also express our heartfelt thanks to the supporters and organizers of the annual conferences over the years, especially Professor Yong He and his colleagues from Xinjiang Normal University, who organized the 2023 meeting, as well as to all our colleagues in the mathematical community.

I hope that readers will find both rewards and joy in this volume of collected papers. I wish the research and the nurturing of talent in geometric analysis in China to reach even greater heights.

Gang Tian

June 2024

Contents

Critical Allard Regularity in Dimension Two and Its Application

YUCHEN BI

Beijing International Center for Mathematical Research, Peking University, Beijing, China

E-mail: ycbi@bicmr.pku.edu.cn

JIE ZHOU

School of Mathematical Sciences, Capital Normal University, Beijing, China

E-mail: zhoujiemath@cnu.edu.cn

Abstract   This is a survey of the results in [2, 3] regarding the bi-Lipschitz regularity of two varifolds with critical Allard condition, and related applications. For an integral 2-varifold $V = \underline{v}(\Sigma, \theta_{\geqslant 1})$ in the unit ball B_1 passing through the origin, assuming the critical Allard condition holds, that is, the area $\mu_V(B_1)$ is close to the area of a unit disk and the generalized mean curvature has sufficient small L^2 norm, we prove Σ is bi-Lipschitz homeomorphic to a flat disk in $\mathbb{R}^2$ locally. For related applications, we discuss the bi-Lipschitz almost rigidity for L^2-almost CMC surfaces.

1   Background

Let $V = \underline{v}(\Sigma, \theta)$ be a rectifiable n-varifold in $\mathbb{R}^{n+k}$ and $\mu = \theta \mathcal{H}^n \llcorner \Sigma$ be the corresponding Radon measure. The generalized mean curvature of the varifold is defined by a function $H \in L^1_{loc}(d\mu)$ satisfying

$$\delta V(X) = \int div^{\Sigma} X d\mu = -\int H \cdot X d\mu, \qquad \forall X \in C^1_c.$$

One of the most fundamental theorems in geometric measure theory is the Allard regularity theorem [1].

Theorem 1 (Allard [1])   *Assume $V = \underline{v}(\Sigma, \theta)$ is a rectifiable n-varifold in $B_1(0)$ such that $0 \in spt\mu, \theta(x) \geqslant 1$ for μ-a.e.x and*

$$\mu(B_1(0)) \leqslant (1+\delta)\omega_n \quad and \quad \left(\int_{B_1(0)} |H|^p d\mu\right)^{\frac{1}{p}} \leqslant \delta$$

for some $p > n$ and $\delta \in (0, \delta_0(n, k, p)]$. Then, after rotation,

$$spt\mu \cap B_\gamma(0) = graph\ u = \{(x, u(x)) | x \in B_\gamma^n(0)\}$$

for some $u \in C^{1,\alpha=1-\frac{n}{p}}$ and $\gamma = \gamma(n, k, p) \ll 1$.

It is an ϵ-regularity theorem on the mean curvature equation in the supercritical case $p > n$. In this presentation, we care about the critical case in dimension two, i.e., $p = n = 2$. We call the following condition as Critical Allard condition:

$$\mu(B_1(0)) \leqslant (1+\delta)\omega_n \quad and \quad \int_{B_1(0)} |H|^n d\mu \leqslant \delta. \tag{1.1}$$

By comparing with the linear equation, the lowest regularity we can expect is the Hölder regularity. The Reifenberg topological disk theorem [26] is a classical criteria for the Hölder regularity of closed subsets in the Euclidean space, which says: if Σ is a closed subset in $B_1(0)$ and for any $x \in \Sigma \cap B_{\frac{1}{2}}(0)$ and $r \in \left(0, \frac{1}{2}\right)$, there exists an n-dimensional affine plane $T_{x,r}$ passing through x such that,

$$\frac{1}{r} d_{\mathcal{H}}(\Sigma \cap B_r(x), T_{x,r} \cap B_r(x)) \leqslant \varepsilon, \tag{1.2}$$

then there exists bi-Hölder map $\varphi : D_{\frac{1}{4}}^n \to \Sigma \cap B_{\frac{1}{4}}(0)$ such that $\varphi(0) = 0$ and

$$(1 - \psi)|x - y|^{2-\alpha} \leqslant |\varphi(x) - \varphi(y)| \leqslant (1 + \psi)|x - y|^\alpha,$$

where $\alpha = \alpha(\varepsilon) \to 1$ and $\psi = \psi(\varepsilon) \to 0$ as $\varepsilon \to 0$.

In Allard's original work [1], he has proved that the critical Allard condition (1.1) implies the Reifenberg condition (1.2) on a fixed scale $r = 1$ and fixed base point $x = 0$. So, to get the C^α regularity under the critical Allard condition, it is enough to show that the Ahlfors regularity condition

$$1 - \gamma \leqslant \frac{\mu(B_r(x))}{\omega_n r^n} \leqslant 1 + \gamma, \quad \forall x \in \Sigma \cap B_1(0), r < 1. \tag{1.3}$$

For rectifiable varifold with critical Allard condition, the Ahlfors regularity is true in two cases. One is by the monotonicity in dimension two under the assumption $H \perp T\Sigma$ [31], the other is by the interplay with the tilt-excess estimate

$$E(x, r, T_{x,r}) := r^{-n} \int_{B(x,r)} |p_{T_y\Sigma} - p_{T_{x,r}}|^2 d\mu \leqslant \gamma^2 \qquad (1.4)$$

during the iteration process under the integral assumption(by [20–22] and private communication). Let us remark that the generalized mean curvature of an integral varifold is perpendicular almost everywhere.

So, for integral varifold satisfying the critical Allard condition (1.1), we know that the Reifenberg condition (1.2), the Ahlfors regularity (1.3) and the tilt-excess estimate (1.4) hold for any $x \in \Sigma$ and $r < 1$ for the same time and the C^α regularity holds. These three items mean the varifold is close to a plane in the sense of Hausdorff distance, in the sense of area density and in the sense that the BMO norm of the Gauss map is small respectively. Moreover, motivated by Semmes' work [27–29] on the $W^{1,p}$ parameterization theorem of chord-arc hypersurfaces in $\mathbb{R}^{n+1}$, we conclude the three conditions and define the conception of chord-arc varifold.

Definition 2 (Chord-arc varifolds)   A rectifiable varifold $V = \underline{v}(\Sigma, \theta)$ is called a chord-arc varifold in $B(x, r)$ with constant γ, if for any $B(\xi, \sigma) \subset B(x, r)$ with $\xi \in \Sigma$ there exists $T_{\xi,\sigma} \in G(n, 2)$ such that (1.2), (1.3) and (1.4) hold for $B(\xi, \sigma)$, i.e.,

$$1 - \gamma \leqslant \frac{\mu(B(\xi, \sigma))}{\pi\sigma^2} \leqslant 1 + \gamma,$$

$$\frac{1}{\sigma} d_{\mathcal{H}}(B(\xi, \sigma) \cap \Sigma, B(\xi, \sigma) \cap (T_{\xi,\sigma} + \xi)) \leqslant \gamma,$$

$$\sigma^{-2} \int_{B(\xi,\sigma)} |p_{T_x\Sigma} - p_{T_{\xi,\sigma}}|^2 d\mu \leqslant \gamma^2.$$

By modifying Semmes' argument, we extend the $W^{1,p}$-parameterization theorem to the non-smooth and higher-codimension case, see Theorem 9. Since $W^{1,p} \subset C^\alpha$ and $\alpha \to 1$ as $p \to \infty$, this $W^{1,p}$ parameterization theorem is a generalization of the C^α regularity. It is a subtle question to ask whether the bi-Hölder and bi-$W^{1,p}$ parameterization can be extended to be a bi-Lipschitz one. On the one hand, if we replace the p in Allard's $C^{1,\alpha=1-\frac{n}{p}}$ regularity theorem by $p = n$, then we will get $\alpha = 0$ and then expect the varifold to have C^1 or $C^{0,1} = W^{1,\infty}$ regularity; on the other hand, even for the linear Poisson equation, when the right hand term is just in L^n, we can only expect the solution to have $W^{2,n}$ regularity, which is precisely not Lipschitz($W^{1,\infty}$). In dimension two, there are some experiences from a geometric viewpoint. That is T.Toro [30] and Müller-Šverák's [24] bi-Lipschitz regularity theorem of surfaces with small total curvature.

Theorem 3 (T.Toro [30], Müller-Sverák [24]) *If $\Sigma \subset \mathbb{R}^n$ is a smooth complete surface with $\int_\Sigma |A|^2 d\mu_g \leqslant \varepsilon$, then there exists conformal parameterization $f :$ $\mathbb{C} \to \Sigma$ s.t. $f^*g = e^{2w}g_{\mathbb{R}^2}$ satisfies*

$$\|w\|_{C^0(\mathbb{C})} + \|\nabla w\|_{L^2(\mathbb{C})} + \|\nabla^2 w\|_{L^1} \leqslant \psi(\varepsilon).$$

The above global statement dues to Müller and Sverák, and Toro's original theorem is a local one which says a surface with bounded area and small total curvature can be locally decomposed to be finite many pieces and each of the pieces is bi-Lipschitz to a disk with small Lipschitz constant. Inspired by these bi-Lipschitz estimates, it is expected that there holds the bi-Lipschitz regularity under the critical Allard condition.

2   Main result and related application

We state our results in two situations. The first one is the a priori bi-Lipschitz estimate for smooth surfaces satisfying the critical Allard condition.

Theorem 4(Bi&Zhou [3]) *For any $\varepsilon, \gamma > 0$, there exists $\delta = \delta(\varepsilon), \sigma = \sigma(\gamma, \varepsilon) >$ 0 such that if $F : (\Sigma, p) \to (B_1, 0)$ is a proper immersion with $F(p) = 0$ satisfying*

$$\mu_g(\Sigma) \leqslant 2\pi(1 - \gamma) \text{ and } \int_\Sigma |H|^2 d\mu_g \leqslant \delta,$$

then there exists a bi-Lipschitz homeomorphism $\varphi : D_\sigma \to U(p) \sim \Sigma^\sigma(p)$ such that $\varphi^*g = e^{2w}g_{\mathbb{R}^2}$ with

$$\|w\|_{C^0(D_\sigma)} + \|Dw\|_{L^2(D_\sigma)} \leqslant \varepsilon.$$

and

$$1 - \varepsilon \leqslant \frac{|\varphi(x) - \varphi(y)|}{|x - y|} \leqslant 1 + \varepsilon.$$

We remark that here, we only need to assume $\mu_g(\Sigma) \leqslant 2\pi(1 - \gamma)$ for some $\gamma > 0$ and do not need to assume that $\mu_g(\Sigma) \leqslant \pi(1 + \delta)$. But respectively, we only have the estimate in a small scale $B_\sigma(0)$ for σ depends on γ to exclude the scaled catenoid. In the non-smooth setting, the statement is as follows.

Theorem 5 (Bi & Zhou [2]) *Assume $V = \underline{v}(\Sigma, \theta)$ is an integral 2-varifold in B_1 such that $0 \in spt\mu_V$ and $\theta \geqslant 1$. If*

$$\mu(B_1) \leqslant (1 + \varepsilon)\pi \quad and \quad \int_{B_1} |H|^2 d\mu \leqslant \varepsilon,$$

Then, there is a bi-Lipschitz conformal map $f : D_1 \to \Sigma$ such that $\Sigma \cap B(0, 1 - \psi) \subset f(D_1)$ and

1. $df \otimes df = e^{2w} g_{\mathbb{R}^2}$ for some $w \in W^{1,2}(D_1) \cap L^\infty(D_1)$ with

$$\|w\|_{L^\infty(D_1)} + \|Dw\|_{L^2(D_1)} \leqslant \psi;$$

2. $f \in W^{2,2}(D_1, \mathbb{R}^n)$ with $\|f - id\|_{W^{2,2}(D_1)} \leqslant \psi$.
Here $\psi = \psi(\varepsilon) \to 0$ as $\varepsilon \to 0$.

Next, we turn to a related application on the almost rigidity of surfaces whose mean curvature is almost a constant. There are many results [4–8,10,12] on this topic. Most of them assume that the mean curvature is close to a constant in the L^∞ sense. And we care about the case in which the mean curvature is close to a constant in some integral sense(especially in the L^n-sense, since this is the critical case). In this direction, Delgadino, Maggi, Mihaila and Neumayer [10] established a result in the general setting of anisotropic crystals. A special corollary of their results is the isotropic version.

Theorem 6 (Delgadino, Maggi, Mihaila & Neumayer [10]) *Assume $\Omega_i \subset \mathbb{R}^{n+1}$ is a sequence of domains such that $H_0(\Omega_i) := \dfrac{n\mathcal{H}^n(\partial\Omega_i)}{(n+1)\mathcal{H}^{n+1}(\Omega_i)} \equiv n$ and*

1. $\exists L \in \mathbb{N}$ and $\sigma, \kappa \in (0, 1)$ such that

$$sup_i diam(\Omega_i) < +\infty,$$

$$\sup_i \mathcal{H}^n(\partial\Omega_i) \leqslant (L + \sigma)\mathcal{H}^n(S^n),$$

$$H_{\partial\Omega_i} \geqslant \kappa n(uniformly\ mean\ convex);$$

2. $\delta_2(\Omega_i) := \left(\fint_{\partial\Omega_i} \left| \frac{H}{H_0} - 1 \right|^2 d\mathcal{H}^n \right)^{\frac{1}{2}} \to 0.$

Then $\Omega_i \to \Omega = \cup_{j=1}^{l \leqslant L} B^n(x_i) = $ union of disjoint unit balls in the sense

$$\lim_{i \to \infty} \left(|\mathcal{H}^n(\partial\Omega_i) - \mathcal{H}^n(\partial\Omega)| + \mathcal{H}^{n+1}(\Omega_i \Delta \Omega) \right) = 0.$$

This result holds for any dimension and only requires the mean curvature to be close to a constant in the L^2 sense. Different from the case the mean curvature is close to a constant in the L^∞ sense, this time they need to assume the uniform mean convexity as a condition. The theorem also deals with the case in which there are many bubbles (which is singular at the touching points) as the limit model. The almost rigidity is in the sense of Hausdorff convergence and flat topology. We want to know whether the uniform mean convex assumption $H_{\partial\Omega_i} \geqslant \kappa$ can be removed and whether the convergence can be stronger. We get a positive answer in dimension two for the case of one bubble.

Theorem 7 (Bi & Zhou [3]) *For any $\alpha \in \left(0, \dfrac{1}{2} \right)$, there exists $\varepsilon = \varepsilon(\alpha) > 0$ such that if $F : \Sigma \to \mathbb{R}^3$ is an immersion of a closed surface with*

$$\int_\Sigma |H - \bar{H}|^2 d\mu_g \leqslant \varepsilon \ \text{and} \ \int_\Sigma |H|^2 d\mu_g \leqslant 32\pi(1 - \alpha),$$

where $H = \langle \vec{H}, \vec{N} \rangle$ and $\bar{H} = \dfrac{1}{\mu_g(\Sigma)} \displaystyle\int_\Sigma H d\mu_g$, then

$$\Sigma \approx S^2 \text{(homeomorphic)}.$$

Moreover, after scaling such that $Area(\Sigma) = 4\pi$, there exists $\bar{F} : S^2 \to F(\Sigma) \subset \mathbb{R}^3$ such that $d\bar{F} \otimes d\bar{F} = e^{2u} g_{S^2}$ with

$$\|\bar{F} - id_{S^2}\|_{W^{2,2}} + \|u\|_{L^\infty} \leqslant \psi(\varepsilon) \to 0 \ \text{as} \ \varepsilon \to 0.$$

Let us explain the role of the condition $\displaystyle\int_\Sigma |H|^2 d\mu_g \leqslant 32\pi(1-\alpha)$. Intuitively, it will be used to rule out the occurrence of two touched bubbles as a limit model, since the least Willmore energy $\dfrac{1}{4} \displaystyle\int_\Sigma |H|^2 d\mu_g$ of a sphere is 4π. More precisely, by the monotonicity formula, the global condition $\int_{\Sigma} |H|^2 d\mu_g \leqslant 32\pi(1-\alpha)$ can be localized to the local density condition

$$\frac{\mathcal{H}^2(\Sigma \cap B_r(p))}{\pi r^2} \leqslant 2(1-\gamma),$$

which combined with the L^2-almost CMC condition implies the critical Allard condition

$$\int_{\Sigma \cap B_a(p)} |H|^2 d\mu_g \leqslant \varepsilon \text{ and } \frac{\mathcal{H}^2(\Sigma \cap B_r(p))}{\pi r^2} \leqslant 2(1-\gamma).$$

This lies in the setting of our a priori estimate and then Theorem 4 applies.

3   Sketch of the ideas

In this section, we sketch the ideas of proving Theorem 5. Let us first recall Müller and Šverák's proof of the bi-Lipschitz regularity of surfaces with a small total curvature. When writing in conformal coordinates, the Gauss curvature equation reads as

$$-\Delta w = Ke^{2w}.$$

Generally speaking, $A \in L^2$ only implies $Ke^{2w} \in L^1$, which is not enough to get $w \in C^0$. But there are some compensations coming from the structure of the Gauss curvature equation. More precisely, when the total curvature $\int_{\Sigma} |A|^2 d\mu_g$ is small, by choosing a Coulomb gauge, the Gauss map $G : D \to \Sigma \to G_2(\mathbb{R}^n) \subset \mathbb{CP}^{n-1}$ can be lifted along the Hopf fibration $\pi : S^{2n-1} \to \mathbb{CP}^{n-1}$ to be a map $\varphi \in W^{1,2}(D, S^{2n-1} \subset \mathbb{C}^n)$ with small energy. What is more important is the Gauss curvature equation can then be written as

$$-\Delta w = Ke^{2w} = \sum_i *(d\varphi^i \wedge d\varphi^i) \in \mathcal{H}^1,$$

and the algebraic structure of $d\varphi^i \wedge d\varphi^i$ implies the right hand side term is not only in L^1, but also in $\mathcal{H}^1$-the Hardy space [9], which is dual to(by [11]) the space of BMO functions. But in dimension two, the fundamental solution of the Laplace equation belongs to BMO exactly. So, such duality will imply the

C^0 estimate of w, which implies the induced metric $g = e^{2w}g_0$ is bi-Lipschitz equivalent to the Euclidean metric g_0.

Now, in our setting, we only know $H \in L^2$. So, we care about the mean curvature equation

$$\Delta F = He^{2w}.$$

Noting $\int |H|^2 d\mu = \int |He^{2w}|^2 e^{-2w} = \int |\Delta F|^2 e^{-2w}$, so we choose $\eta = e^{-2w}$ and hope to apply the weighted L^2 estimate: if η is an A_2-weight, then weighted L^2-estimate implies

$$\int |A|^2 d\mu \leqslant \int |D^2 F|^2 \eta dx \lesssim \int |\Delta F|^2 \eta dx = \int |H|^2 d\mu,$$

and hence we come back to the situation of small total curvature. Here a function h is called a Muchenhoupt A_p-weight [23] if

$$\fint_{D(x,r)} h \cdot \left(\fint_{D(x,r)} h^{-\frac{1}{p-1}} \right)^{p-1} \leqslant C, \quad \forall x, r > 0.$$

Especially, when $p = 2$, the John-Nirenberg inequality implies if $w \in BMO$, then $\eta = e^{-2w}$ is an A_2-weight.

So, we are facing two difficulties: the existence of conformal coordinates in the non-smooth setting and the BMO estimates of the conformal factor.

To get the existence of conformal coordinate for varifold with critical Allard condition in dimension two, we will apply Lythack and Wenger's theory [15–19] on the solution of Plateau problem in metric spaces.

Theorem 8(Lytchak and Wenger) *If $X \approx D$ is a metric space admits quadratic isoperimetric inequality, then, there exists a harmonic map $f \in C^0(D, X) \cap W_{loc}^{1,p>2}(D)$ such that*

1. if the approximating metric derivative $Df = ap\ mdf_z$ is induced by inner product, then f is conformal,

2. if X is bi-Lipschitz to geodesic space, then $f : D \to X$ is a homeomorphism.

By item 1, for the harmonic map to be conformal, we restrict the Euclidean metric on the varifold. Reifenberg's theorem implies the varifold is homeomorphic to a disk. To show the harmonic map is a parametrization, by item 2, we need to show the restricted metric is bi-Lipschitz to the geodesic metric. That is, the varifold should satisfy the chord-arc property. In the smooth setting, Semmes studied [27–29] the relationship between the chord-arc properties and the tilt-excess estimates and Reifenberg conditions of a hypersurface. Especially, Semmes proved [28] that such hypersurfaces admit $W^{1,p}$ parametrization. Motivated by this, we conclude the Reifenberg condition, the Ahlfors regularity and the tilt-excess estimate to introduce the conception of chord-arc varifold (See Definition 2) and modified Semmes' $W^{1,p}$ parametrization theorem to the non-smooth and higher-codimension situation.

Theorem 9 (Semmes [28], Bi&Zhou [2]) *Assume* $V = \underline{v}(\Sigma, \theta)$ *is a chord-arc varifold with constant* $\gamma \ll 1$. *Then, for any* $\xi \in \Sigma, \sigma > 0$, *there exists a* $W^{1,p}$ *map* $f_{\xi,\sigma} : D_\sigma \to \Sigma$ *such that*

  1. $|f_{\xi,\sigma}(x) - x - \xi| \leqslant C\psi\sigma;$

  2. $B(\xi, (1 - \psi)\sigma) \cap \Sigma \subset f_{\xi,\sigma}(D_\sigma) \subset B(\xi, \sigma) \cap \Sigma$

  3. $\displaystyle\int_{D_\sigma} |D(f_{\xi,\sigma} - i_{\xi,\sigma})|^p \leqslant C_p\psi\sigma^2,$

where $D_\sigma \subset T_{\xi,\sigma}$ *and* $i_{\xi,\sigma}(x) = x$, $\displaystyle\lim_{\gamma \to 0} \psi(\gamma) \to 0$.

We remark that for two dimensional varifold with critical Allard condition, a $W^{1,p}$ parametrization results can also be obtained by verifying the Carleson condition for Jones' β-number in Naber-Valtorta's $W^{1,p}$-reifenberg theorem [25]. With this theorem, we show that any chord-arc varifold really has the chord-arc property.

Proposition 10 (Bi & Zhou [2]) *Assume* $V = \underline{v}(\Sigma, \theta)$ *is a chord-arc varifold with constant* $\gamma \ll 1$, *then for any* $\xi \in \Sigma, \sigma > 0$ *there exists* $\Omega_{\xi,\sigma} \subset \Sigma$ *such that,*

  1. $B(\xi, (1 - \psi)\sigma) \cap \Sigma \subset \Omega_{\xi,\sigma} \subset B(\xi, \sigma) \cap \Sigma;$

  2. there exists $C = C(\gamma_0, \lambda), C_1 = C_1(\gamma_0, \lambda, \Omega_{\xi,\sigma})$ *such that*

$$\frac{1}{C_1}d_{\xi,\sigma}(x, y) \leqslant \frac{1}{C}d(x, y) \leqslant |x - y| \leqslant d(x, y) \leqslant d_{\xi,\sigma}(x, y);$$

  3. $(\Omega_{\xi,\sigma}, |\cdot|)$ *admits quadratic isoperimetric inequality.*

Since varifolds satisfying the critical Allard condition are all chord-arc varifolds, the above proposition allows us to apply Lytchak and Wenger's theory(Theorem 8) to show the existence of conformal coordinates on such varifolds.

So, it remains to estimate the BMO norm of the conformal factor. For this, a key ingredient is Heinonen and Koskela's theory for quasi-conformal mappings [13, 14]. More precisely, Heinonen and Koskela's work implies the conformal parametrization $f : D \to \Omega_{0,\lambda}$ we find is quasi-symmetry:

$$\frac{|f(y) - f(x)|}{|f(z) - f(x)|} \leqslant \eta \left(\frac{|y - x|}{|z - x|} \right), \quad \forall y, z \in D(x, r) \subset D(x, 8r) \subset D_1,$$

where $\eta(t) = 0$ as $t \to 0$. This is a scaling invariant estimate, which allows us to apply the blowup argument to show that the map can be approximated by linear conformal maps.

Theorem 11 (Bi&Zhou [2]) *Assume $V = \underline{v}(\Sigma, \theta)$ is a chord-arc varifold in B_1 with constant $\gamma \ll 1$ with $0 \in \Sigma$. $f : D \to \Omega_\lambda$ =the conformal parameterization from Lytchak and Wenger. Then, for any $D(x, r) \subset D_{1-\psi}$ there exists an affine map $A_{x,r}(y) = a_{x,r} T_{x,r} y + b_{x,r}$ such that*

$$\frac{1}{a_{x,r} r} \left(\|f(y) - A_{x,r}(y)\|_{C^0(D(x,r))} + \|\nabla(f - A_{x,r})\|_{L^2(D(x,r))} \right) \leqslant \psi,$$

where $a_{x,r} > 0$, $b_{x,r} \in \mathbb{R}^n$, and $T_{x,r}$ is a linear isometry from $\mathbb{R}^2$ into $\mathbb{R}^n$.

Geometrically, this means the area of the image of any square is close to (dominated by) the square of the length of its edge. Analytically, it implies the inverse Hölder inequality with small constant: for any x, r with $Q_{x,r} \subset D_{1-\psi}$,

$$\fint_{Q_{x,r}} |det \nabla f| \leqslant (1 + \psi) \left(\fint_{Q_{x,r}} |det \nabla f|^{\frac{1}{2}} \right)^2,$$

which implies the BMO estimate of the conformal factor

$$\|w = \log |det \nabla f|\|_{BMO(Q_{x,r})} \leqslant \psi, \quad \forall x, r \text{ with } Q_{x,r} \subset D_{1-\psi}.$$

Thus by the John-Nirenberg inequality, we know

$$\fint_{Q_{x,r}} e^{2w} dx \fint_{Q_{x,r}} e^{-2w} dx \leqslant C,$$

and hence $\eta = e^{-2w}$ is an A_2-weight. So, by applying the weighted elliptic estimate to the mean curvature equation, we come back to the situation of finite total curvature.

4 Questions

To conclude, let me ask two questions parallel to our topic. One is for higher dimension and the other is for intrinsic geometry.

- Question A. How about the higher dimensional case?

$$\int_{B_1(0) \cap \Sigma^n} |A|^n d\mu \leqslant \varepsilon \Rightarrow \text{ Lipschitz regularity?}$$

- Question B. How about the intrinsic case?

$$\int_{B_1(p)} |Ric^-|^{\frac{n}{2}} dx \leqslant \varepsilon \text{ and } vol(B_1(p)) \geqslant (1-\varepsilon)\omega_n$$

$$\Rightarrow |\frac{vol(B_r(x))}{\omega_n r^n} - 1| \leqslant \delta, \quad \forall x, \forall r?$$

References

[1] W. K. Allard, *On the first variation of a varifold*, Ann. Math., 95(3): 417–491, 1972.

[2] Y. C. Bi, J. Zhou, *Bi-Lipschitz regularity of 2-varifolds with the critical Allard condition*, arXiv:2212.03043, 2022.

[3] Y. C. Bi, J. Zhou, *Bi-Lipschitz rigidity for L^2-almost CMC surfaces*, arXiv:2212.02946, 2022.

[4] G. Ciraolo, A. Figalli, F. Maggi, et al., *Rigidity and sharp stability estimates for hypersurfaces with constant and almost-constant nonlocal mean curvature*, J. Reine Angew. Math., 741: 275–294, 2018.

[5] G. Ciraolo, F. Maggi, *On the shape of compact hypersurfaces with almost-constant mean curvature*, Comm. Pure Appl. Math., 70(4): 665–716, 2017.

[6] G. Ciraolo, A. Roncoroni, L. Vezzoni, *Quantitative stability for hypersurfaces with almost constant curvature in space forms*, Ann. Mat. Pura Appl., 200: 2043–2083, 2021.

[7] G. Ciraolo, L. Vezzoni, *A sharp quantitative version of Alexandrov's theorem via the method of moving planes*, J. Eur. Math. Soc., 20(2): 261–299, 2018.

[8] G. Ciraolo, L. Vezzoni, *Quantitative stability for hypersurfaces with almost constant mean curvature in the hyperbolic space*, Indiana Univ. Math. J., 69(4): 1105–1153, 2020.

[9] R.Coifman, P. L. Lions, Y. Meyer, et al., *Compensated compactness and Hardy spaces*, J. Math. Pures Appl., 72(9): 247–286, 1993.

[10] M. G. Delgadino, F. Maggi, C. Mihaila, et al., *Bubbling with L^2-almost constant mean curvature and an Alexandrov-type theorem for crystals*, Arch. Ration. Mech. Anal., 230(3): 1131–1177, 2018.

[11] C. Fefferman, E. M. Stein, *$\mathcal{H}^p$ spaces of several variables*, Acta Math., 129(3-4): 137–193, 1972.

[12] M. G. Delgadino, F. Maggi, *Alexandrov's theorem revisited*, Anal. PDE, 12(6): 1613–1642, 2019.

[13] J. M. Heinonen, P. Koskela, *Quasiconformal maps in metric spaces with controlled geometry*, Acta Mathematica, 181: 1–61, 1998.

[14] J. M. Heinonen, P. Koskela, N.Shanmugalingam, et al., *Sobolev Spaces on Metric Measure Spaces*, Cambridge University Press, 2015.

[15] A. Lytchak, S. Wenger, *Regularity of harmonic discs in spaces with quadratic isoperimetric inequality*, Calc. Var. Partial Differential Equations, 55: 1–19, 2015.

[16] A. Lytchak, S. Wenger, *Area minimizing discs in metric spaces*, Arch. Ration. Mech. Anal., 223: 1123–1182, 2015.

[17] A. Lytchak, S. Wenger, *Energy and area minimizers in metric spaces*, Adv. Calc. Var., 10(4): 407–421, 2017.

[18] A. Lytchak, S. Wenger, *Intrinsic structure of minimal discs in metric spaces*, Geom. Topol., 22(1): 591–644, 2018.

[19] A. Lytchak, S. Wenger, *Canonical parameterizations of metric disks*, Duke Math. J., 169(4): 761–797, 2020.

[20] U. Menne, *Some applications of the isoperimetric inequality for integral varifolds.*, Adv. Calc. Var., 2(3): 247–269, 2009.

[21] U. Menne, *A Sobolev Poincaré type inequality for integral varifolds.*, Calc. Var. Partial Differential Equations, 38(3-4): 369–408, 2010.

[22] U. Menne, *Decay estimates for the quadratic tilt-excess of integral varifolds*, Arch. Ration. Mech. Anal., 201(1): 1–83, 2012.

[23] B. Muckenhoupt, *Weighted norm inequalities for the Hardy maximal function*, Trans. Ame. Math. Soc., 165: 207–226, 1972.

[24] S. Müller, V. Šverák, *On surfaces of finite total curvature*, J. Differential Geom., 42(2): 229–258, 1995.

[25] A. Naber, D. Valtorta, *Rectifiable-Reifenberg and the regularity of stationary and minimizing harmonic maps*, Ann. Math., 185(1): 131–227, 2017.

[26] E.R. Reifenberg, *Solution of the Plateau Problem for m-dimensional surfaces of varying topological type*, Acta Math., 104: 1–92, 1960.

[27] S. Semmes, *Chord-arc surfaces with small constant, I*, Adv. Math., 85(2): 198–223, 1991.

[28] S. Semmes, *Chord-arc surfaces with small constant, II*, Adv. Math., 88(2): 170–199, 1991.

[29] S. Semmes, *Hypersurfaces in $\mathbb{R}^n$ whose unit normal has small BMO norm*, Proc. Amer. Math. Soc., 112(2): 403–412, 1991.

[30] T.Toro, *Surfaces with generalized second fundamental form in L^2 are Lipschitz manifolds*, J. Differential Geom., 39(1): 65–101, 1994.

[31] J.Zhou, *Topology of surfaces with finite Willmore energy*, Int. Math. Res. Not., 2022(9): 7100–7151, 2022.

A Survey on the Complex Hessian Equation on Compact Kähler Manifolds

JIANCHUN CHU

School of Mathematical Sciences, Peking University, Beijing, China

E-mail: jianchunchu@math.pku.edu.cn

Abstract We survey some recent works on the complex Hessian equation on compact Kähler manifolds.

1 Introduction

The fully non-linear elliptic PDEs play a significant role in the study of differential geometry, since many problems can be transformed into PDEs. There are deep connections between the existence of solutions of PDEs and geometric properties of the underlying manifold.

One of the important geometric PDEs is the complex Hessian equation. Let (M, ω) be a compact Kähler manifold of complex dimension n and $1 \leqslant k \leqslant n$ be a positive integer. For a function φ on M, we write

$$\omega_\varphi = \omega + \sqrt{-1}\partial\bar{\partial}\varphi.$$

Then the complex Hessian equation can be written as

$$
\begin{cases}
\omega_\varphi^k \wedge \omega^{n-k} = f\omega^n, \\
\omega_\varphi \in \Gamma_k(M, \omega), \\
\sup_M \varphi = 0,
\end{cases}
\tag{1.1}
$$

where f is a non-negative function on M satisfying the compatibility condition

$$\int_M f\omega^n = \int_M \omega^n \tag{1.2}$$

and $\Gamma_k(M,\omega)$ denotes the k-th Gårding cone of (M,ω) (see Subsection 2.1).

The equation (1.1) covers two important PDEs. When $k = 1$, (1.1) is the Laplacian equation. When $k = n$, (1.1) is the complex Monge-Ampère equation, which is very related to Calabi's conjecture (see [3]). Yau [18] solved Calabi's conjecture by proving the existence of solutions to the complex Monge-Ampère equation. This result is known as the Calabi-Yau theorem and occupies a crucial place in the study Kähler geometry. Hence, (1.1) $(2 \leqslant k \leqslant n-1)$ can be regarded as a natural interpolation between the Laplacian equation and the complex Monge-Ampère equation.

In this survey, we only focus on the case when $2 \leqslant k \leqslant n-1$. For the topic of the complex Monge-Ampère equation, we refer the reader to Phong-Song-Sturm's survey [15] and the references therein.

This survey is organized as follows. In Section 2, we review some necessary definitions and results. In Sections 3 and 4, we describe some recent results of the non-degenerate and degenerate complex Hessian equation on compact Kähler manifolds respectively, including the existence, uniqueness and regularity of solutions.

2 Preliminaries

2.1 The k-th Gårding cone

Let σ_k and Γ_k be the k-th elementary symmetric polynomial and the Gårding cone in $\mathbb{R}^n$, i.e.

$$\sigma_k(\lambda) = \sum_{i_1 < \cdots < i_k} \lambda_{i_1} \cdots \lambda_{i_k} \quad \text{for } \lambda = (\lambda_1, \cdots, \lambda_n),$$

and

$$\Gamma_k = \{\lambda \in \mathbb{R}^n \mid \sigma_i(\lambda) > 0 \text{ for } i = 1, \cdots, k\}.$$

We use $A^{1,1}(M)$ to denote the space of all smooth $(1,1)$-forms on M. For any $\alpha \in A^{1,1}(M)$, let $\lambda(\alpha)$ be the eigenvalues of α with respect to ω. It then follows that

$$\sigma_k(\lambda(\alpha)) = \binom{n}{k} \frac{\alpha^k \wedge \omega^{n-k}}{\omega^n}, \quad \text{where} \quad \binom{n}{k} = \frac{n!}{k!(n-k)!}.$$

The corresponding k-th Gårding cone of (M,ω) is defined by

$$\Gamma_k(M,\omega) = \{\alpha \in A^{1,1}(M) \mid \sigma_i(\lambda(\alpha)) > 0 \text{ for } i = 1,\cdots,k\}.$$

2.2 k-subharmonic functions and maximality

Definition 1 Let Ω be a domain in $\mathbb{C}^n$ and β be the standard Euclidean metric, i.e.

$$\beta = \sum_k \sqrt{-1}dz_i \wedge d\bar{z}_i.$$

An upper semi-continuous function $\varphi : \Omega \to [-\infty, +\infty)$ is called k-subharmonic if $\varphi \in L^1_{\text{loc}}(\Omega)$ is subharmonic and for any $\alpha_1, \ldots, \alpha_{k-1} \in \Gamma_k(\Omega, \beta)$, the following inequality holds in the sense of currents:

$$\sqrt{-1}\partial\bar{\partial}\varphi \wedge \alpha_1 \wedge \ldots \wedge \alpha_{k-1} \wedge \omega^{n-k} \geqslant 0 \text{ in } \Omega.$$

Denote by $\text{SH}_k(\Omega)$ the set of all k-subharmonic functions on Ω.

Definition 2 The function $\varphi \in \text{SH}_k(\Omega)$ is called maximal if the following holds. For any $\psi \in \text{SH}_k(\Omega)$ and any $K \subset \Omega$, if $\varphi \geqslant \psi$ on $\Omega \setminus K$, then $\varphi \geqslant \psi$ on K.

Błocki [2] gave a characterization of maximality for bounded k-subharmonic functions.

Proposition 3 (Błocki [2]) *Suppose that $\varphi \in \text{SH}_k(\Omega)$ is bounded. Then φ is maximal if and only if the following equality holds in the sense of measures:*

$$(\sqrt{-1}\partial\bar{\partial}\varphi)^k \wedge \omega^{n-k} = 0 \text{ in } \Omega.$$

2.3 (ω, k)-subharmonic functions and (ω, k)-polar sets

Definition 4 Let Ω be a domain in $\mathbb{C}^n$ and ω be a positive definite $(1,1)$-form in Ω. A function $\varphi : \Omega \to [-\infty, +\infty)$ is called ω-subharmonic if

    (1) φ is upper semi-continuous and locally integrable;

    (2) $\sqrt{-1}\partial\bar{\partial}\varphi \wedge \omega^{n-1} \geqslant 0$ in the sense of currents;

    (3) if ψ satisfies (1), (2) and $\psi = \varphi$ almost everywhere in Ω, then $\varphi \leqslant \psi$.

Definition 5 Let (M,ω) be a compact Kähler manifold. An upper semi-continuous function $\varphi : M \to [-\infty, +\infty)$ is called (ω, k)-subharmonic if

    (1) in any coordinate chart $U \subset M$, $\rho + \varphi$ is ω-subharmonic in U, where ρ denotes the local potential of ω, i.e. $\omega = \sqrt{-1}\partial\bar{\partial}\rho$ in U;

(2) for any $\alpha_1, \cdots, \alpha_{k-1} \in \Gamma_k(M, \omega)$, the following inequality holds in the sense of currents:

$$(\omega + \sqrt{-1}\partial\bar{\partial}\varphi) \wedge \alpha_1 \wedge \ldots \wedge \alpha_{k-1} \wedge \omega^{n-k} \geqslant 0.$$

Denote by $\mathrm{SH}_k(M, \omega)$ the set of all (ω, k)-subharmonic functions on M.

Definition 6   For $\varphi \in \mathrm{SH}_k(M, \omega)$, set $\varphi_i = \max(\varphi, -i)$ for any i and define a measure $\mathrm{H}_k(\varphi)$ by

$$\mathrm{H}_k(\varphi) = \lim_{i \to \infty} \left(\mathbf{1}_{\{\varphi > -i\}} \omega_{\varphi_i}^k \wedge \omega^{n-k} \right).$$

Denote by $\mathcal{E}(M, \omega, k)$ the set of all (ω, k)-subharmonic functions on M such that

$$\int_M \mathrm{H}_k(\varphi) = \int_M \omega^n.$$

Definition 7   A Borel set $E \subset M$ is called (ω, k)-polar if there exists $\varphi \in \mathrm{SH}_k(M, \omega)$ such that $E \subset \{\varphi = -\infty\}$.

2.4   Gauduchon metric and Laplacian equation

Definition 8   Let M be a complex manifold of complex dimension n. A Hermitian metric ω_G is said to be Gauduchon if

$$\partial\bar{\partial}\omega_G^{n-1} = 0.$$

Gauduchon [7] showed that every Hermitian metric is conformal to a Gauduchon metric.

Theorem 9 (Gauduchon [7])   *Let $(M, \hat{\omega})$ be a compact Hermitian manifold of complex dimension n. Then there exists $\hat{u} \in C^\infty(M)$, unique up to the addition of a constant, such that the conformal metric $e^{\hat{u}}\hat{\omega}$ is Gauduchon, i.e.*

$$\sqrt{-1}\partial\bar{\partial}(e^{\hat{u}}\hat{\omega})^{n-1} = 0.$$

Definition 10   Let $(M, \hat{\omega})$ be a compact Hermitian manifold of complex dimension n. The canonical Laplacian operator $\Delta_{\hat{\omega}}$ is defined as

$$\Delta_{\hat{\omega}} v = \frac{n\sqrt{-1}\partial\bar{\partial}v \wedge \omega^{n-1}}{\omega^n} \quad \text{for } v \in C^2(M).$$

The following theorem proved by Gauduchon [7] gives the equivalent characterization of the solvability for the Laplacian equation on compact Hermitian manifold.

Theorem 11 (Gauduchon [7]) *Let $(M, \hat{\omega})$ be a compact Hermitian manifold of complex dimension n. Fix a non-negative integer l, and $\alpha \in (0, 1)$. Given $h \in C^{l,\alpha}(M)$, there exists $v \in C^{l+2,\alpha}(M)$ solving*

$$\begin{cases} \Delta_{\hat{\omega}} v = h, \\ \sup_{M} v = 0 \end{cases}$$

if and only if

$$\int_{M} h e^{(n-1)\hat{u}} \hat{\omega}^n = 0,$$

where $\hat{u}$ is the function in Theorem 9.

3 Non-degenerate case

The complex Hessian equation (1.1) is said to be non-degenerate if the right-hand function f is positive. Such equation was studied under some assumptions on (M, ω) (see [9,11,12]). For the general case, thanks to the a priori estimates of Hou [9], Hou-Ma-Wu [10] and Dinew-Kołodziej [5], the non-degenerate complex Hessian equation was solved by the continuity method.

Theorem 12 (Hou [9], Hou-Ma-Wu [10], Dinew-Kołodziej [5]) *Let (M, ω) be a compact Kähler manifold and $f \in C^{\infty}(M)$ be a positive function satisfying the compatibility condition (1.2). Then there exists a unique $\varphi \in C^{\infty}(M)$ solving (1.1)*

Next, we will discuss the arguments of zeroth order estimate and complex Hessian estimate, including original and alternative proofs.

3.1 Zeroth order estimate

There are four methods to establish the zeroth order estimate. The first method is to follow Yau's Moser iteration method [18] for the complex Monge-Ampère equation.

Theorem 13 (Hou [9]) *Let (M, ω) be a compact Kähler manifold and $f \in C^\infty(M)$ be a positive function satisfying the compatibility condition* (1.2). *If $\varphi \in C^\infty(M)$ is a solution of* (1.1), *then for any $p > 2n$, there exists a constant C depending only on $\|f\|_{L^p}$, p and (M, ω) such that*

$$\|\varphi\|_{L^\infty} \leqslant C.$$

The second method is to generalize Kołodziej's pluripotential theory [13] for the complex Monge-Ampère equation.

Definition 14 Let (M, ω) be a compact Kähler manifold of complex dimension n. For any Borel set $E \subset M$, the Hessian k-capacity of E is defined as

$$\mathrm{Cap}_{\omega,k}(E) = \sup \left\{ \int_E \omega_\varphi^k \wedge \omega^{n-k} \mid \varphi \in \mathrm{SH}_k(M, \omega),\ 0 \leqslant \varphi \leqslant 1 \right\}.$$

Theorem 15 (Dinew-Kołodziej [4]) *Let (M, ω) be a compact Kähler manifold and $f \in C^\infty(M)$ be a positive function satisfying the compatibility condition* (1.2). *If $\varphi \in C^\infty(M)$ is a solution of* (1.1), *then for any $p > \dfrac{n}{k}$, there exists a constant C depending only on $\|f\|_{L^p}$, p and (M, ω) such that*

$$\|\varphi\|_{L^\infty} \leqslant C.$$

By using the Alexandrov-Bakelman-Pucci maximum principle, Błocki [1] established the zeroth order estimate for the complex Monge-Ampère equation. Székelyhidi [16] generalized this approach to a general class of fully non-linear elliptic equations on compact Hermitian manifolds, including the complex Hessian equation on compact Kähler manifolds.

Theorem 16 (Székelyhidi [16]) *Let (M, ω) be a compact Kähler manifold and $f \in C^\infty(M)$ be a positive function satisfying the compatibility condition* (1.2). *If $\varphi \in C^\infty(M)$ is a solution of* (1.1), *then there exists a constant C depending only on $\|f\|_{L^\infty}$ and (M, ω) such that*

$$\|\varphi\|_{L^\infty} \leqslant C.$$

Using Wang-Wang-Zhou's idea [17], Guo-Phong-Tong [8] gave a pure PDE proof of zeroth order estimate of a general class of fully non-linear elliptic equations on compact Kähler manifolds, including the complex Hessian equation.

Theorem 17 (Guo-Phong-Tong [8]) *Let (M, ω) be a compact Kähler manifold and $f \in C^\infty(M)$ be a positive function satisfying the compatibility condition (1.2). If $\varphi \in C^\infty(M)$ is a solution of (1.1), then for any $p > n$, there exists a constant C depending only on $\|f\|_{L^{\frac{n}{k}}(\log L)^p}$, p and (M, ω) such that*

$$\|\varphi\|_{L^\infty} \leqslant C.$$

3.2 Complex Hessian estimate

Theorem 18 (Hou-Ma-Wu [10], Dinew-Kołodziej [5]) *Let (M, ω) be a compact Kähler manifold and $f \in C^\infty(M)$ be a positive function satisfying the compatibility condition (1.2). If $\varphi \in C^\infty(M)$ is a solution of (1.1), then there exists a constant C depending only on $\|f^{\frac{1}{k}}\|_{C^2}$ and (M, ω) such that*

$$\sup_M |\partial\bar{\partial}\varphi| \leqslant C.$$

The above complex Hessian estimate was proved by blow-up analysis. There are two important ingredients: Hou-Ma-Wu's estimate [10] and Dinew-Kołodziej's Liouville theorem [5].

Theorem 19 (Hou-Ma-Wu [10]) *Let (M, ω) be a compact Kähler manifold and $f \in C^\infty(M)$ be a positive function satisfying the compatibility condition (1.2). If $\varphi \in C^\infty(M)$ is a solution of (1.1), then there exists a constant C depending only on $\|f^{\frac{1}{k}}\|_{C^2}$ and (M, ω) such that*

$$\sup_M |\partial\bar{\partial}\varphi| \leqslant C \sup_M |\partial\varphi|^2 + C.$$

Theorem 20 (Dinew-Kołodziej [5]) *If φ is a bounded and maximal k-subharmonic function on $\mathbb{C}^n$ with bounded gradient, then φ is constant.*

For the reader's convenience, we show how to use blow-up analysis to prove Theorem 18.

Proof of Theorem 18 Thanks to Theorem 19, it suffices to establish the gradient estimate. We argue by contradiction. Suppose that the gradient estimate fails, then there exists a sequence of smooth functions $\{\varphi_i\}_{i=1}^\infty$ such that

(1) φ_i solves the following equation

$$\begin{cases} \omega_{\varphi_i}^k \wedge \omega^{n-k} = f_i \omega^n, \\ \omega_{\varphi_i} \in \Gamma_k(M, \omega), \\ \sup_M \varphi_i = 0, \end{cases}$$

for some positive smooth function f_i on M satisfying the compatibility condition

$$\int_M f_i \omega^n = \int_M \omega^n;$$

(2) $\|f_i^{\frac{1}{k}}\|_{C^2} \leqslant C$ for a uniform constant C independent of i;

(3) $L_i^2 := \sup_M |\partial \varphi_i|^2 \to \infty$ as $i \to \infty$.

For each i, let p_i be the maximum point of $|\partial \varphi_i|^2$. After passing to a subsequence, we assume that $p_i \to p_0$ as $i \to \infty$. Choose the holomorphic coordinate system $(B_2(0), \{z_i\}_{i=1}^n)$ centered at p_∞ such that

$$\omega(p_\infty) = \omega(0) = \beta,$$

where β denotes the standard Euclidean metric, i.e.

$$\beta = \sqrt{-1} \sum_k dz_k \wedge d\bar{z}_k.$$

Define (1,1)-form $\hat{\omega}_i$ and function $\hat{\varphi}_i$ on $B_{L_i}(0)$ by

$$\hat{\omega}_i(z) = \omega\left(p_i + \frac{z}{L_i}\right) \text{ and } \hat{\varphi}_i(z) = \varphi_i\left(p_i + \frac{z}{L_i}\right).$$

Then we obtain

$$\sup_{B_{L_i}(0)} |\hat{\varphi}_i| + \sup_{B_{L_i}(0)} |\partial \hat{\varphi}_i| + \sup_{B_{L_i}(0)} |\partial \bar{\partial} \hat{\varphi}_i| \leqslant C$$

and

$$\lim_{i \to \infty} |\partial \hat{\varphi}_i(0)| = \lim_{i \to \infty} \frac{|\partial \varphi_i(p_i)|}{L_i} = 1.$$

After passing to a subsequence, we assume that

(a) $\hat{\varphi}_i \to \hat{\varphi}_\infty$ in $C_{\text{loc}}^{1,\gamma}(\mathbb{C}^n)$ for some $\hat{\varphi}_\infty$;

(b) $|\hat{\varphi}_\infty| + |\partial \hat{\varphi}_\infty| \leqslant C$ and $|\partial \hat{\varphi}_\infty(0)| = 1$.

We claim that $\hat{\varphi}_\infty$ is a maximal k-subharmonic function on $\mathbb{C}^n$. Given claim and by Theorem 20, $\hat{\varphi}_\infty$ is constant, which contradicts with (b). Next, we split the proof of claim into two steps.

Step 1. $\hat{\varphi}_\infty$ is a k-subharmonic function on $\mathbb{C}^n$.

It suffices to prove that $\hat{\varphi}_\infty$ is a k-subharmonic function in $B_R(0)$ for any R. Since $\hat{\omega}_i \to \beta$ in $C^0(B_R(0))$, then for any $\hat{\alpha}_j \in \Gamma_k(B_R(0), \beta)$ and $\varepsilon > 0$, when i is sufficiently large, we have

$$\hat{\alpha}_{j,\varepsilon} := \hat{\alpha}_j + \varepsilon\beta \in \Gamma_k(B_R(0), \hat{\omega}_i).$$

Define

$$\alpha_{j,\varepsilon,i}(z) = \hat{\alpha}_{j,\varepsilon}(L_i(z - p_i))$$

and so

$$\hat{\alpha}_{j,\varepsilon}(z) = \alpha_{j,\varepsilon,i}\left(p_i + \frac{z}{L_i}\right).$$

It then follows that

$$\left[\frac{1}{L_i^2}\hat{\omega}_i + \sqrt{-1}\partial\bar{\partial}\hat{\varphi}_i\right] \wedge \hat{\alpha}_{1,\varepsilon} \wedge \ldots \wedge \hat{\alpha}_{k-1,\varepsilon} \wedge \hat{\omega}_i^{n-k}(z)$$

$$= \frac{1}{L_i^2}(\omega + \sqrt{-1}\partial\bar{\partial}\varphi_i) \wedge \alpha_{1,\varepsilon,i} \wedge \ldots \wedge \alpha_{k-1,\varepsilon,i} \wedge \omega^{n-k}\left(p_i + \frac{z}{L_i}\right) > 0.$$

Letting $i \to \infty$ and then $\varepsilon \to 0$, we obtain

$$\sqrt{-1}\partial\bar{\partial}\hat{\varphi}_\infty \wedge \hat{\alpha}_1 \wedge \ldots \wedge \hat{\alpha}_{k-1} \wedge \beta^{n-k} \geqslant 0.$$

This shows that $\hat{\varphi}_\infty \in \mathrm{SH}_k(B_R(0), \beta)$.

Step 2. $\hat{\varphi}_\infty$ is maximal.

By (1) and the definition of $\hat{\varphi}_i$, we compute

$$\left(\frac{1}{L_i^2}\hat{\omega}_i + \sqrt{-1}\partial\bar{\partial}\hat{\varphi}_i\right)^k \wedge \hat{\omega}_i^{n-k}(z)$$

$$= \frac{1}{L_i^{2k}}(\omega + \sqrt{-1}\partial\bar{\partial}\varphi_i)^k \wedge \omega^{n-k}\left(p_i + \frac{z}{L_i}\right)$$

$$= \frac{1}{L_i^{2k}}f_i\,\omega^n\left(p_i + \frac{z}{L_i}\right).$$

Letting $i \to \infty$, we obtain

$$\sqrt{-1}\partial\bar{\partial}\hat{\varphi}_\infty^k \wedge \beta^{n-k} = 0.$$

This implies the maximality of $\hat{\varphi}_\infty$ by Theorem 3.   □

3.3 The continuity method

Given the a priori estimates in Subsections 3.1 and 3.2, Theorem 12 can be proved by the continuity method. For the reader's convenience, we give all the details here.

Proof of Theorem 12 The uniqueness of solutions follows from the strong maximum principle. Next, we apply the continuity method to prove the existence of solutions. For $t \in [0, 1]$, we consider the family of equations:

$$\begin{cases} \omega_{\varphi_t}^k \wedge \omega^{n-k} = f^t e^{b_t} \omega^n, \\ \omega_{\varphi_t} \in \Gamma_k(M, \omega), \\ \sup_M \varphi_t = 0. \end{cases} \tag{3.1}$$

Fix a constant $\alpha \in (0, 1)$ and define

$$I = \{t \in [0, 1] \mid (3.1) \text{ admits a solution } (\varphi_t, b_t) \in C^{4,\alpha}(M) \times \mathbb{R}\}.$$

It is clear that $0 \in I$. It suffices to show that I is open and closed. Given this, we obtain $1 \in I$ and so there exists $\varphi \in C^{4,\alpha}(M)$ solving (1.1). The standard elliptic theory shows $\varphi \in C^\infty(M)$.

Step 1. I is open.

Define the map $F : C^{4,\alpha}(M) \times \mathbb{R} \to C^{2,\alpha}(M)$ by

$$F(\varphi, b) = \log \frac{\omega_\varphi^k \wedge \omega^{n-k}}{\omega^n} - b.$$

For any $\hat{t} \in I$, let $\hat{L}$ be the linearized operator of $\log \dfrac{\omega_\varphi^k \wedge \omega^{n-k}}{\omega^n}$ at $\varphi_{\hat{t}}$. Since $\hat{L}$ is an elliptic operator without zeroth and first order term, then $\hat{L}$ is given by the canonical Laplacian of some Hermitian metric $\hat{\omega}$ on M, i.e.

$$\hat{L} = \Delta_{\hat{\omega}},$$

and so the linearized operator of F can be expressed by

$$DF|_{(\varphi_{\hat{t}}, b_{\hat{t}})}(v, c) = \Delta_{\hat{\omega}} v - c \quad \text{for } (v, c) \in C^{4,\alpha}(M) \times \mathbb{R}.$$

For any $h \in C^{2,\alpha}(M)$, there exists $c \in \mathbb{R}$ such that

$$\int_M (h + c)e^{(n-1)\hat{u}} \hat{\omega}^n = 0.$$

Then Theorem 11 shows there exists $v \in C^{4,\alpha}(M)$ such that

$$\begin{cases} \Delta_{\hat{\omega}} v = h + c, \\ \sup_M v = 0, \end{cases}$$

which implies

$$DF|_{(\varphi_{\hat{t}}, b_{\hat{t}})}(v, c) = h,$$

and so $DF|_{(\varphi_{\hat{t}}, b_{\hat{t}})}$ is surjective. For the injectivity of $DF|_{(\varphi_{\hat{t}}, b_{\hat{t}})}$, suppose that $(v_1, c_1), (v_2, c_2) \in C^{4,\alpha}(M) \times \mathbb{R}$ satisfy

$$\Delta_{\hat{\omega}} v_1 - c_1 = \Delta_{\hat{\omega}} v_2 - c_2.$$

It then follows that

$$\Delta_{\hat{\omega}}(v_1 - v_2) = c_1 - c_2.$$

Applying the maximum principle at the extrema of $v_1 - v_2$, we obtain $c_1 = c_2$ and then the strong maximum principle shows $v_1 = v_2$.

Step 2. I is closed.

For any $t \in I$, there exists a pair (φ_t, b_t) solving (3.1). Applying the maximum principle at the extrema of φ_t,

$$|b_t| \leqslant \sup_M |\log f|. \tag{3.2}$$

By a priori estimates in Subsections 3.1 and 3.2, Evans-Krylov theory and the bootstrapping method, we see that

$$\|\varphi_t\|_{C^{4,\alpha}} \leqslant C \tag{3.3}$$

for some constant C independent of t. Combining (3.2), (3.3) and the Arzelà-Ascoli theorem, I is closed. $\qquad \square$

4 Degenerate case

The complex Hessian equation (1.1) is said to be degenerate if the right-hand function f is non-negative and vanishes somewhere. There are some results for the degenerate complex Hessian equation even when the right-hand of (1.1) is just a positive Radon measure.

Theorem 21(Dinew-Kołodziej [4]) *Let (M, ω) be a compact Kähler manifold of complex dimension n and f be a non-negative function satisfying $f \in L^p(M, \omega)$ for some $p > \dfrac{n}{k}$ and the compatibility condition (1.2). Then there exists a unique $\varphi \in \mathrm{SH}_k(M, \omega) \cap C(M)$ solving (1.1).*

Theorem 22 (Lu-Nguyen [14]) *Let (M, ω) be a compact Kähler manifold of complex dimension n and μ be a positive Radon measure on M satisfying the compatibility condition*

$$\mu(M) = \int_M \omega^n$$

Assume that μ does not charge m-polar subsets of M. Then there exists a $\varphi \in \mathcal{E}(M, \omega, k)$ solving

$$\omega_\varphi^k \wedge \omega^{n-k} = \mu.$$

Dinew-Pliś-Zhang [6] improved Theorem 19 in the non-degenerate setting by replacing $\|f^{\frac{1}{k}}\|_{C^2}$ by $\|f^{\frac{1}{k-1}}\|_{C^2}$ as follows.

Theorem 23 (Dinew-Pliś-Zhang [6]) *Let (M, ω) be a compact Kähler manifold and $f \in C^\infty(M)$ be a positive function satisfying the compatibility condition (1.2). If $\varphi \in C^\infty(M)$ is a solution of (1.1), then there exists a constant C depending only on $\|f^{\frac{1}{k-1}}\|_{C^2}$ and (M, ω) such that*

$$\sup_M |\partial\bar\partial\varphi| \leqslant C \sup_M |\partial\varphi|^2 + C.$$

Combining Theorem 23 with Theorem 20 and blow-up analysis in Subsection 3.2, one obtains the complex Hessian estimate. Since this estimate is independent of the positive lower bound of the right-hand side f, it has the following corollary in the degenerate setting.

Corollary 24(Dinew-Pliś-Zhang [6]) *Let (M, ω) be a compact Kähler manifold of complex dimension n and f be a non-negative function satisfying $f^{\frac{1}{k-1}} \in$*

$C^{1,1}(M,\omega)$ and the compatibility condition (1.2). If $\varphi \in \text{SH}_k(M,\omega)$ solves (1.1), then φ is weak $C^{1,1}$, i.e. the complex Hessian of φ is bounded.

References

[1] Błocki Z. *On uniform estimate in Calabi-Yau theorem,* Sci. China Ser. A, 48 (2005), 244–247.

[2] Błocki Z. *Weak solutions to the complex Hessian equation,* Ann. Inst. Fourier (Grenoble), 55 (2005), no. 5, 1735–1756.

[3] Calabi E. *On Kähler manifolds with vanishing canonical class*//Algebraic Geometry and Topology. A symposium in honor of S. Lefschetz, 78-89. Princeton: Princeton University Press, 1957.

[4] Dinew S, Kołodziej S. *A priori estimates for complex Hessian equations,* Anal. PDE, 7 (2014), no. 1, 227–244.

[5] Dinew S, Kołodziej S. *Liouville and Calabi-Yau type theorems for complex Hessian equations,* Amer. J. Math., 139 (2017), no. 2, 403–415.

[6] Dinew S, Pliś S, Zhang X. *Regularity of degenerate Hessian equations,* Calc. Var. Partial Differential Equations, 58 (2019), no. 4, Paper No. 138, 21 pp.

[7] Gauduchon P. *Le théorème de l'excentricité nulle,* C. R. Acad. Sci. Paris Sér. A-B, 285 (1977), no. 5, A387–A390.

[8] Guo B, Phong D H, Tong F. *On L^∞ estimates for complex Monge-Ampère equations,* Ann. of Math., (2) 198 (2023), no. 1, 393–418.

[9] Hou Z. *Complex Hessian equation on Kähler manifold,* Int. Math. Res. Not. IMRN, 2009(2009), no. 16, 3098–3111.

[10] Hou Z, Ma X N, Wu D. *A second order estimate for complex Hessian equations on a compact Kähler manifold,* Math. Res. Lett., 17 (2010), no. 3, 547–561.

[11] Jbilou A. *Équations hessiennes complexes sur des variétés kählériennes compactes,* C. R. Math. Acad. Sci. Paris, 348 (2010), no. 1-2, 41–46.

[12] Kokarev V N. *Mixed volume forms and a complex equation of Monge-Ampère type on Kähler manifolds of positive curvature,* Izv. Ross. Akad. Nauk Ser. Mat., 74 (2010), no. 3, 65–78.

[13] Kołodziej S. *The complex Monge-Ampère equation,* Acta Math., 180 (1998), no. 1, 69–117.

[14] Lu C H, Nguyen V D. *Degenerate complex Hessian equations on compact Kähler manifolds,* Indiana Univ. Math. J., 64 (2015), no. 6, 1721–1745.

[15] Phong D H, Song J, Sturm J. *Complex Monge-Ampère equations*//Surveys in Differential Geometry. Vol. XVII, 327-410. Boston: International Press, 2012.

[16] Székelyhidi G. *Fully non-linear elliptic equations on compact Hermitian manifolds,* J. Differential Geom., 109 (2018), no. 2, 337–378.

[17] Wang J, Wang X J, Zhou B. *A priori estimate for the complex Monge-Ampère equation,* Peking Math. J., 4 (2021), no. 1, 143–157.

[18] Yau S T. *On the Ricci curvature of a compact Kähler manifold and the complex Monge-Ampère equation, I,* Comm. Pure Appl. Math., 31 (1978), no. 3, 339–411.

Minimal Graphs over Manifolds

QI DING

Shanghai Center for Mathematical Sciences, Fudan University, Shanghai, China

E-mail: dingqi@fudan.edu.cn

Abstract This is a survey of results in [26, 28, 31, 32] regarding geometry and analysis of minimal graphs over manifolds. We consider rigidity, regularity and asymptotic structure of minimal graphs. We will study Liouville type theorems, Harnack estimates, gradient estimates and splitting of minimal graphs by establishing Neumann-Poincaré inequality on minimal graphs, and also relates this to the Half-space properties of minimal hypersurfaces and asymptotic estimates of minimal graphs by capacities.

1 Motivation

Our results touch minimal graphs in Euclidean space, area-minimizing hypersurfaces in manifolds, and harmonic functions on manifolds. All of these are objects of long history and there are numerous of papers on them.

One important motivation in studying minimal graphs in Euclidean space is the famous Bernstein conjecture: it was asked whether an entire minimal graph in $\mathbb{R}^{n+1}$ has to be an n-plane. The Bernstein conjecture (theorem) was achieved by successive efforts of Fleming [40], De Giorgi [25], Almgren [1] and Simons [76] up to dimension seven within the framework of geometric measure theory. On the other hand, Bombieri De Giorgi-Giusti [7] in 1969 then provided a counterexample by constructing a non-trivial entire minimal graph in $\mathbb{R}^{n+1}$ with $n > 7$. Furthermore, Simon studied the existence of entire minimal graphs in $\mathbb{R}^{n+1}(n > 7)$ with prescribed cylindrical tangent cones at infinity [74], and the uniqueness of tangent cones of entire minimal graphs in Euclidean space [75] (see also [78]).

In Euclidean space, area-minimizing hypersurfaces have been studied intensely for several decades before (see [42] [60] [73] for a systematical introduction). The theory acts an important role in the Bernstein theorem. In particular, minimal graphs in Euclidean space are automatically area-minimizing. For any area-minimizing hypersurface M in $\mathbb{R}^{n+1}$ (or a smooth manifold), De Giorgi [24], Federer [36], Reifenberg [66] proved that the singular set of M has Hausdorff dimension $\leqslant n - 7$. Recently, Cheeger-Naber [16] and Naber-Valtorta [65] made important progress on quantitative stratifications of the singular set of stationary varifolds. Needless to say, all area-minimizing hypersurfaces in smooth manifolds are stable. The theory of stable minimal surfaces is a powerful tool to study the topology of 3-dimensional manifolds, see [2] [39] [61] [70] [71] for instance.

The equation for harmonic functions is the linear analogue of the minimal hypersurface equation on Riemannian manifolds. Harmonic functions on complete manifolds with non-negative Ricci curvature have been very successfully studied by Yau [80], Cheeger-Colding [11], Colding-Minicozzi [20], Li [54] and many others. In particular, Green functions on complete manifolds with non-negative Ricci curvature admit stronger properties, see Colding [18], Colding-Minicozzi [19], Li-Tam-Wang [56] and so on. Some properties of harmonic functions can be extended to manifolds of volume doubling property and $(1, 1)$-type Poincaré inequality (see for instance [20]). On the other hand, the theory of harmonic functions is a powerful tool to study the structure of manifolds of Ricci curvature bounded below, see [11–14, 17] for instance.

2   Interior gradient estimates

Let Σ be an n-dimensional complete Riemannian manifold with Levi-Civita connection D. Let Ω be an open subset of Σ, and $M = \{(x, u(x)) \mid x \in \Omega\}$ be a graph over Ω in $\Sigma \times \mathbb{R}$. M is said to be *a minimal graph* (over Ω) if M is a minimal hypersurface in $\Sigma \times \mathbb{R}$, or equivalently u satisfies

$$\mathrm{div}_\Sigma \left(\frac{Du}{\sqrt{1 + |Du|^2}} \right) = 0 \tag{2.1}$$

on Ω. Here, div_Σ denotes the divergence of Σ. This equation is called the minimal hypersurface equation, and u is said to be the *minimal graphic function*, which is a critical point of the volume functional of graphs

$$\int_\Omega \sqrt{1 + |Df|^2}.$$

This is also equivalent to that the height function is harmonic on M. Namely,

$$\Delta_M u = 0,$$

where we have identified $u(x) = u(x, u(x))$ for each $x \in \Omega$, and Δ_M denotes the Laplacian of M w.r.t. its induced metric from $\Sigma \times \mathbb{R}$. Moreover, M is area-minimizing in $\Omega \times \mathbb{R}$.

In [7], Bombieri-De Giorgi-Giusti proved the gradient estimate for every smooth solution u to minimal hypersurface equation (2.1) on a Euclidean ball $B_r(x)$: there are two constants c_1, c_2 so that

$$|Du(x)| \leqslant c_1 e^{c_2 r^{-1}\left(\sup_{B_r(x)} u - u(x)\right)}, \tag{2.2}$$

where Finn got the 2-dimensional case in [37], and the estimate (2.2) is sharp in the sense of the linear exponential dependence on the solutions from Finn's example [38]. Moreover, Korevaar [52] and Wang [79] showed gradient estimates for minimal hypersurface equation by the maximum principle, and Spruck [77] got it for the case of manifolds. In [67], Rosenberg-Schulze-Spruck proved an interior gradient estimate for minimal graphs over manifolds of nonnegative Ricci curvature, where the estimation depends on the lower bound of sectional curvature.

3　Liouville type theorems

In the 1840s, Liouville proved that any bounded holomorphic function on $\mathbb{C}$ is constant. In 1975, Yau [80] proved that every positive harmonic function on a complete manifold Σ of nonnegative Ricci curvature, $Ric \geqslant 0$, is a constant.

From (2.2), any entire positive minimal graphic function on $\mathbb{R}^n$ is a constant (see [7]). It is natural to ask *Liouville type theorem* for positive minimal graphic functions on complete manifolds with $Ric \geqslant 0$. The 'positive' condition is necessary from our examples in [26].

For two dimensions, all positive minimal graphic functions on complete manifolds with nonnegative sectional curvature are constant from the classification by Fischer-Colbrie and Schoen [39]. In 2013, Rosenberg-Schulze-Spruck [67] proved that any positive minimal graphic function on a complete manifold of $Ric \geqslant 0$ is a constant provided the manifold has sectional curvature uniformly bounded from below. Casteras-Heinonen-Holopainen [10] proved that any positive minimal graphic function u on a complete manifold of asymptotically nonnegative sectional curvature is a constant provided u has at most linear growth.

Theorem 1 (Liouville type theorem [28]) *Any positive minimal graphic function on a complete manifold of $Ric \geqslant 0$ is a constant.*

In fact, the above result is valid in a more general setting as follows. Let Σ be a complete manifold satisfying *volume doubling property*

$$\text{vol}(B_{2r}(x)) \leqslant c \, \text{vol}(B_r(x)), \tag{3.1}$$

and *(1,1)-Poincaré inequality*

$$\int_{B_r(x)} |f - f_{B_r(x)}| \leqslant cr \int_{B_r(x)} |Df| \tag{3.2}$$

for all $x \in \Sigma$ and $r > 0$. Here, $f_{B_r(x)} = \dfrac{1}{\text{vol}(B_r(x))} \displaystyle\int_{B_r(x)} f$, and c is an absolute constant. Given $p \in \Sigma$, let $\bar{p} = (p, u(p))$ and $B_R(\bar{p})$ be the geodesic ball in $\Sigma \times \mathbb{R}$ centered at $\bar{p}$ with the radius R.

Theorem 2 (Harnack's inequality [28]) *Let M be a minimal graph over $B_{4R}(p)$ $\subset \Sigma$ with the graphic function $u > 0$. Then u satisfies*

$$\sup_{B_R(\bar{p}) \cap M} u \leqslant \vartheta \inf_{B_R(\bar{p}) \cap M} u. \tag{3.3}$$

The key ingredient in the proof of Theorem 2 is to establish the Sobolev inequality and the Neumann-Poincaré inequality for the positive monotonic C^1-functions of the minimal graphic function u. And this is sufficient to carry out De Giorgi-Nash-Moser iteration for the positive monotonic C^1-functions of u.

As a corollary, we immediately obtain that any positive minimal graphic function on Σ is a constant from Theorem 2.

Remark 1 *Shortly after us [28], Colombo-Magliaro-Mari-Rigoli [21] also* *proved Theorem 1 with a different technique, which depends on the lower bound of Ricci curvature.*

If we suppose stronger conditions on base manifolds, then it is possible to get Liouville type theorems for minimal graphs without positive graphic functions. We suppose that Σ is a complete n-dimensional Riemannian manifold of nonnegative Ricci curvature and Euclidean volume growth, which has quadratic decay of the curvature tensor.

Theorem 3 ([26]) *Suppose Σ has non-radial Ricci curvature satisfying*

$$\inf_{\partial B_\rho(p)} Ric_\Sigma \left(\xi^T, \xi^T \right) \geqslant \kappa \rho^{-2} |\xi^T|^2 \qquad a.e.$$

for the distance function ρ from p and every tangent vector ξ on Σ, where ξ^T stands for the part that is tangential to the geodesic sphere $\partial B_\rho(p)$. If $\kappa > \dfrac{(n-3)^2}{4}$, then any minimal graphic function on Σ must be constant.

The constant $\dfrac{(n-3)^2}{4}$ in Theorem 3 is sharp from our examples of non-trivial minimal graphs in [26]. The proof of Theorem 3 uses stability inequality on minimal graph M over Σ and that any tangent cone of M at infinity is asymptotically vertical at infinity, which is proved by estimating the measure of a 'bad' set and stability arguments (as in [27]) to eliminate the unbounded situation.

4   Sobolev inequality and Neumann-Poincaré inequality

Let $B_{2R}(p)$ be an n-dimensional geodesic ball centered at p with radius $2R$ and with smooth metric, whose Ricci curvature $\geqslant -(n-1)\kappa^2 R^{-2}$ for some $\kappa \geqslant 0$ and $R \geqslant 1$. From Anderson [3] or Croke [23] (see also [45,57] for instance), there is a constant $\alpha_{n,\kappa} > 0$ depending only on n, κ such that the Sobolev inequality holds

$$\frac{\alpha_{n,\kappa}}{R} \left(\text{vol}(B_R(p))\right)^{\frac{1}{n}} \left(\int_{B_R(p)} |\phi|^{\frac{n}{n-1}} \right)^{\frac{n-1}{n}} \leqslant \int_{B_R(p)} |D\phi| \qquad (4.1)$$

for any Lipschitz function ϕ on $B_R(p)$ with compact support in $B_R(p)$, where D is the Levi-Civita connection of $B_{2R}(p)$. From Buser [9] or Cheeger-Colding [11], the Neumann-Poincaré inequality holds on $B_R(p)$. Namely, up to choosing the constant $\alpha_{n,\kappa} > 0$, it holds

$$\alpha_{n,\kappa} \int_{B_R(p)} |f - f_{B_R(p)}| \leqslant R \int_{B_R(p)} |Df| \tag{4.2}$$

for any Lipschitz function f on $B_R(p)$ with $f_{B_R(p)} = \dfrac{1}{\text{vol}(B_R(p))} \displaystyle\int_{B_R(p)} f$. These two inequalities play an important role in partial differential equation and geometry.

In 1967, Miranda obtained a Sobolev inequality for minimal graphs in Euclidean space [64]. After that, Bombieri [4] and Michael (see the appendix of [72]) gave a simpler proof of this inequality. More general, Michael-Simon [63] proved the Sobolev inequality on arbitrary submanifolds in Euclidean space.

The Sobolev inequality on minimal submanifolds can be generalized from Euclidean space to Riemannian manifolds. In [48], Hoffman-Spruck obtained the Sobolev inequality on a minimal submanifold M in a manifold N, with some geometric restrictions involving the volume of M, the sectional curvatures of N and the injectivity radius of N (see [49] [68] for more results). Recently, Brendle [8] proved the Sobolev inequality on minimal submanifolds in manifolds of nonnegative sectional curvature and Euclidean volume growth.

In 1972, Bombieri-Giusti [6] proved the celebrated uniform Neumann-Poincare inequality on any area-minimizing hypersurface M in $\mathbb{R}^{n+1}$. Namely, there is a constant $c_n \geqslant 1$ depending only on n such that

$$\min_k \left(\int_{B_r(x) \cap M} |f - k|^{\frac{n}{n-1}} \right)^{\frac{n-1}{n}} \leqslant c_n \int_{B_{c_n r}(x) \cap M} |\nabla f| \tag{4.3}$$

for any $x \in M$, $r > 0$, $f \in C^1(M)$. Here, ∇ denotes the Levi-Civita connection on M.

We obtain Sobolev and Neumann-Poincaré inequalities on minimal graphs over manifolds with Ricci curvature bounded below as follows.

Theorem 4 ([31]) *For two constants $\kappa \geqslant 0$ and $v > 0$, let Σ be an n-dimensional smooth complete noncompact manifold with Ricci curvature*

$$\text{Ric} \geqslant -(n-1)\kappa^2 \text{ on } B_2(p), \text{ and } \text{vol}(B_1(p)) \geqslant v. \tag{4.4}$$

There is a constant $\Theta = \Theta_{\kappa,v} \geqslant 1$ depending only on n, κ, v such that if M is a minimal graph over $B_2(p)$ with $\bar{p} = (p, 0) \in M$ and $\partial M \subset \partial B_2(p) \times \mathbb{R}$, then there hold a Sobolev inequality

$$\left(\int_M |\phi|^{\frac{n}{n-1}} \right)^{\frac{n-1}{n}} \leqslant \Theta \int_M |\nabla \phi| \tag{4.5}$$

for any function $\phi \in C_0^1(M \cap B_1(\bar{p}))$, and a Neumann-Poincaré inequality

$$\int_{M \cap B_{1/\Theta}(\bar{p})} |\varphi - \bar{\varphi}| \leqslant \Theta \int_{M \cap B_1(\bar{p})} |\nabla \varphi| \tag{4.6}$$

for any function $\varphi \in C^1(M \cap B_1(\bar{p}))$ with $\bar{\varphi} = \dfrac{1}{\mathrm{vol}(M \cap B_{1/\Theta}(\bar{p}))} \displaystyle\int_{M \cap B_{1/\Theta}(\bar{p})} \varphi$.
Here, ∇ is the Levi-Civita connection with respect to the induced metric on M.

For $M = B_2(p) \times \{0\} \subset \Sigma \times \mathbb{R}$, the inequalities (4.5)(4.6) reduce to (4.1)(4.2) with $R = 1$. Hence, the constant $\Theta_{\kappa,v}$ in Theorem 4 indeed depends on the constants κ, v. Once we get (4.6), then (4.5) can be easily derived combining with the covering technique. The sketch proof of Theorem 4 will be given in section 6.

Moreover, our argument of the proof of Theorem 4 works for area-minimizing hypersurfaces in almost Euclidean balls. In particular, we have a generalization of (4.3) as follows.

Theorem 5 (Neumann-Poincaré inequality [31]) *Let N be an $(n + 1)$-dimensional complete noncompact manifold with $\mathrm{Ric} \geqslant -n\varepsilon^2$ on $B_2(q) \subset N$ and $\mathrm{vol}(B_1(q)) \geqslant (1 - \varepsilon)\omega_n$ for a constant $\varepsilon \in (0, 1)$. Let M be an area-minimizing hypersurface in $B_2(q)$ with $q \in M$ and $\partial M \subset \partial B_2(q)$. For the suitable small $\varepsilon > 0$, there is a constant $\Theta_n > 0$ such that*

$$\inf_k \left(\int_{M \cap B_{1/\Theta_n}(q)} |\varphi - k|^{\frac{n}{n-1}} \right)^{\frac{n-1}{n}} \leqslant \Theta_n \int_{M \cap B_1(q)} |\nabla \varphi| \tag{4.7}$$

for any function $\varphi \in C^1(M \cap B_1(q))$.

Let Σ be an n-dimensional complete manifold with Ricci curvature $\geqslant -(n-1)\kappa^2 r^{-2}$ on the geodesic ball $B_r(p) \subset \Sigma$. Suppose $\mathrm{vol}(B_r(x)) \geqslant vr^n$. Compared with the Euclidean result (2.2) by Bombieri-De Giorgi-Miranda, we [31] showed that the minimal graphic function u on $B_r(p)$ satisfies

$$|Du(x)| \leqslant ce^{cr^{-1}\left(\sup_{B_r(x)} u - u(x)\right)} \tag{4.8}$$

for some constant $c = c(n, \kappa, v)$.

If Σ further has nonnegative Ricci curvature and Euclidean volume growth, then the gradient $|Du|$ can be bounded by a polynomial of u of n-th power by a similar argument of Bombieri-Giusti [6]. In particular, if u has linear growth from one side, then $\sup_{\Sigma} |Du| < \infty$.

Inspired by Cheeger-Colding-Minicozzi [15], we showed that any tangent cone of Σ at infinity splits off a line isometrically whenever there are linear growth minimal graphic functions on Σ in [31]. Here, the 'linear growth' is necessary for splitting from our example in [26].

Shortly after us, Colombo-Gama-Mari-Rigoli [22] studied the splitting of manifolds with nonnegative Ricci curvature and the further assumption that the $(n-2)$-th Ricci curvature in radial direction is bounded below by $Cr^{-2}(x)$, whenever there are linear growth minimal graphic functions on them.

5 Capacity and applications

Let Σ be a complete manifold with the Levi-Civita connection D. Let $\Omega \subset \Sigma$ be an open set, and $K \subset \Sigma$ be a closed set with $K \subset \Omega$ and compact ∂K. For a constant $t \geqslant 0$, let $\mathcal{L}_t(K, \Omega)$ denote the set containing all locally Lipschitz functions ϕ on Σ with compact $\overline{\mathrm{spt}\phi \setminus K}$ in Ω such that $0 \leqslant \phi \leqslant t$ and $\phi\big|_K = t$, and $\mathrm{spt}\phi$ denotes the support of ϕ in Σ.

The classical capacity is given by

$$\mathrm{cap}(K, \Omega) = \inf_{\phi \in \mathcal{L}_1(K, \Omega)} \int |D\phi|^2. \tag{5.1}$$

It's well-known that the infimum is attained by harmonic functions on $\Omega \setminus K$. Now, we define a capacity by

$$\mathrm{cap}_t(K, \Omega) = \inf_{\phi \in \mathcal{L}_t(K, \Omega)} \int \left(\sqrt{1 + |D\phi|^2} - 1 \right). \tag{5.2}$$

It's easy to see that

$$\text{cap}_t(K,\Omega) \leqslant \frac{t^2}{2}\text{cap}(K,\Omega) = \frac{t^2}{2} \inf_{\mathcal{L}_1(K,\Omega)} \int |D\phi|^2, \tag{5.3}$$

and

$$\frac{T}{t}\text{cap}_t(K,\Omega) \leqslant \text{cap}_T(K,\Omega) \leqslant \frac{T^2}{t^2}\text{cap}_t(K,\Omega) \qquad \text{for all } 0 < t \leqslant T.$$

Suppose $\partial\Omega \cup \partial K$ is Lipschitz-continuous. Let φ_t be a function on $\partial\Omega \cup \partial K$ with $\varphi_t = t$ on ∂K and $\varphi = 0$ on $\partial\Omega$. From Giusti [42],

$$\text{cap}_t(K,\Omega) = \inf_{f \in BV(\Omega\backslash K)} \left(\int_{\Omega\backslash K} \left(\sqrt{1+|Df|^2} - 1\right) + \int_{\partial(\Omega\backslash K)} |\text{tr}f - \varphi_t| \right).$$

Moreover, there exists a function $u \in BV(\Omega\setminus K)$ so that u is a smooth solution to (2.1) on $\Omega\setminus K$ and

$$\text{cap}_t(K,\Omega) = \int_{\Omega\backslash K} \left(\sqrt{1+|Du|^2} - 1\right) + \int_{\partial(\Omega\backslash K)} |\text{tr}u - \varphi_t|.$$

Using geometric measure theory, we can study it in a more general setting without the Lipschitz condition on $\partial(\Omega\setminus K)$.

Theorem 6 ([32]) *Let Ω be a bounded open set in Σ, and K be a compact set in Ω with $\partial\overline{\Omega\setminus K} = \partial(\Omega\setminus K)$. For any $t > 0$, there is a function u of bounded variation on Σ with $0 \leqslant u \leqslant t$ such that u is a smooth solution to (2.1) on $\Omega\setminus K$, $u = t$ on $K\setminus\partial K$, $u = 0$ on $\Sigma\setminus\overline{\Omega}$, the boundary of the subgraph $U = \{(x,s) \in \Sigma \times \mathbb{R}|\, s < u(x)\}$ is countably n-rectifiable set in $\Sigma \times \mathbb{R}$, and*

$$\text{cap}_t(K,\Omega) = \int_{\Omega\backslash K} \left(\sqrt{1+|Du|^2} - 1\right) + \int_{\partial K} (t-u) + \int_{\partial\Omega} u.$$

Moreover, $\{x \in \partial\Omega|\, u(x) > 0\}$ and $\{x \in \partial K|\, u(x) < t\}$ are both countably $(n-1)$-rectifiable, u is lower semi-continuous on a small neighborhood of ∂K, and upper semi-continuous on a small neighborhood of $\partial\Omega$.

The function u here is said to be a (BV) solution on Σ associated with $\text{cap}_t(K,\Omega)$. The condition $\partial\overline{\Omega\setminus K} = \partial(\Omega\setminus K)$ is necessary, which is equivalent to that for any $x \in \partial(\Omega\setminus K)$ and any $r > 0$

$$\text{vol}(B_r(x) \cap \Omega\setminus K) < \text{vol}(B_r(x)).$$

Remark 2   *For the case of mean convex $\Omega \setminus K$, the Dirichlet problem for minimal hypersurface equation on $\Omega \setminus K$ is solvable for classic solutions from Jenkins-Serrin [51]. In general, it may not have a classic solution u with the prescribed boundary data $u = t$ on ∂K and $u = 0$ on $\partial \Omega$.*

Proposition 7   *The solution on Σ associated with $\mathrm{cap}_t(K, \Omega)$ is unique up to a constant.*

Lin [59] proved the uniqueness (on minimizing currents) of Dirichlet problem for minimal hypersurface equation on bounded Lipschitz domains of Euclidean space.

Theorem 8   *Let Ω be a bounded open set in Σ, and K be a compact set in Ω. Suppose that $\partial \overline{\Omega \setminus K} = \partial(\Omega \setminus K)$. If an integer multiplicity current T is minimizing in $\overline{\Omega \setminus K} \times \mathbb{R}$ with $T = [|K|] \times \{t\} + [|\Sigma \setminus \Omega|] \times \{0\}$ in $\left(\Sigma \setminus \overline{\Omega \setminus K}\right) \times \mathbb{R}$ for some $t > 0$. Then $\mathrm{spt} T$ is the boundary of the subgraph of a solution u associated with $\mathrm{cap}_t(K, \Omega)$, i.e.,*

$$T = \partial[|\{(x, s) \in \Sigma \times \mathbb{R} \,|\, s < u(x)\}|].$$

Suppose that there is a sequence of open sets $\Omega_i \supset K$ with $\partial \overline{\Omega_i \setminus K} = \partial(\Omega_i \setminus K)$ and $d(p, \partial \Omega_i) \to \infty$ for some $p \in \Sigma$. For each i, let u_i be a BV solution on Σ associated with $\mathrm{cap}_t(K, \Omega_i)$. There holds $u_i \leqslant u_{i+1}$ on Σ. Hence, there is a BV function u on Σ so that $U \cap \partial\{(x, s) \,|\, x \in \Sigma, \, s < u_i(x)\}$ converges as $i \to \infty$ to $U \cap \partial\{(x, s) \,|\, x \in \Sigma, \, s < u(x)\}$ in the Hausdorff sense for any bounded open set $U \subset \Sigma \times \mathbb{R}$. We say that u is a *(BV) solution on Σ associated with* $\mathrm{cap}_t(K) := \mathrm{cap}_t(K, \Sigma)$.

We can prove the uniqueness of functions on Σ associated with $\mathrm{cap}_t(K)$, and $\inf_{\Sigma \setminus K} u < t$ provided $\mathrm{cap}_t(K) > 0$.

Remark 3   *In general, we actually do not have $\inf_{\Sigma \setminus K} u = 0$, and do not have*
$$\lim_{r \to \infty} \mathrm{cap}_t(B_r(p)) = \infty \text{ provided } \mathrm{cap}_t(B_1(p)) > 0 \text{ (see Appendix II in [32]).}$$

If $\mathrm{cap}(K) := \mathrm{cap}(K, \Sigma) = 0$ for some compact $K \subset \Sigma$ with $\mathrm{vol}(K) > 0$, then Σ is said to be *parabolic*. Otherwise, Σ is said to be *nonparabolic*, which is equivalent to that there are positive Green functions on Σ. If there are a compact set $K \subset \Sigma$ with $\mathrm{vol}(K) > 0$ and $t > 0$ so that $\mathrm{cap}_t(K) = 0$, then we call that Σ is *M-parabolic*. Otherwise, we call that Σ is *M-nonparabolic*. In particular, a complete parabolic manifold must be M-parabolic.

Theorem 9 ([32]) *The following properties are equivalent.*

(1) There are a constant $t > 0$ and a compact set K in Σ so that $\mathrm{cap}_t(K) > 0$.

(2) For any $t > 0$ and any compact set K in Σ with $\mathrm{vol}(K) > 0$, there holds $\mathrm{cap}_t(K) > 0$.

(3) For any compact K in Σ with $\mathrm{vol}(K) > 0$, there is a smooth positive non-constant bounded solution u to (2.1) on $\Sigma \setminus K$.

(4) There is a non-flat smooth bounded graph over Σ in $\Sigma \times \mathbb{R}$ with nonnegative mean curvature.

Let us review the history briefly on the 'half-space property'. In 1990, Hoffman-Meeks [47] proved a famous half-space theorem. It asserts that any complete proper immersed minimal surface in $\mathbb{R}^2 \times \mathbb{R}^+$ is $\mathbb{R}^2 \times \{c\}$ for some constant $c \in \mathbb{R}^+ = (0, \infty)$. Let Σ be a complete Riemannian manifold, and M be a complete minimal hypersurface properly immersed in $\Sigma \times \mathbb{R}^+$. We say that Σ has the *half-space property* if M must be a slice $\Sigma \times \{c\}$ for some constant $c \in \mathbb{R}^+$. In 2013, Rosenberg-Schulze-Spruck [67] proved an interesting result that a recurrent manifold with bounded sectional curvature has the half-space property. Here, the recurrence is equivalent to the parabolicity (see Grigor'yan [44]). Recently, Colombo-Magliaro-Mari-Rigoli [21] proved that a complete parabolic manifold with Ricci curvature bounded below has the half-space property.

The half-space property can be seen as a special case of Frankel-type theorems [41] for minimal hypersurfaces in manifolds. This has a close relation to the maximum principle at infinity for minimal hypersurfaces that has been studied in [33], [53], [62] and related references therein.

Using the capacity in (5.2), we can show that a M-parabolic manifold has the half-space property as follows, which infers directly that all complete parabolic manifolds have the half-space property.

Theorem 10 (Half-space property [32]) *Every M-parabolic manifold has the half-space property. Namely, any complete connected minimal hypersurface properly immersed in $\Sigma \times \mathbb{R}^+$ must be a slice $\Sigma \times \{c\}$ for some $c \in \mathbb{R}^+$ provided*

Σ *is a M-parabolic manifold.*

From the proof of Theorem 10, we immediately have the following corollary.

Corollary 11 *For a M-parabolic manifold P, any smooth mean concave domain in $P \times [0, \infty)$ must be $P \times (c, c')$ for constants $0 \leqslant c < c' \leqslant \infty$.*

Our M-parabolic condition in the above Corollary is sharp from (4) of Theorem 9.

Before proving Theorem 10, we need the following lemma.

Lemma 12 *Let K be a compact set in Σ and $\Omega_i \supset K$ be a sequence of open sets with $\partial \overline{\Omega_i \setminus K} = \partial(\Omega_i \setminus K)$ and $d(p, \partial \Omega_i) \to \infty$ for some $p \in \Sigma$. If u_i is a BV solution on Σ associated with $\mathrm{cap}_t(K, \Omega_i)$ for some $t > 0$, then $u_i \to t$ locally on Σ in the C^0-sense.*

Let S be a minimal hypersurface properly immersed in $\Sigma \times \mathbb{R}^+$. By contradiction, assume S is not a slice. By translating S vertically, we can assume that $S \subset \Sigma \times \mathbb{R}^+$ and there is a constant $\delta > 0$ such that $S \cap (\Sigma \times \{\delta'\}) \neq \emptyset$ for any $\delta' \in (0, \delta)$. Hence, there are a point $p \in \Sigma$, a compact set $K \subset B_1(p)$, and a constant $\tau \in (0, \delta)$ such that $S \cap (K \times [0, \tau]) = \emptyset$. For each integer $i \geqslant 1$, let u_i be a BV solution on Σ associated with $\mathrm{cap}_\tau(K, B_i(p))$ for each i. Let

$$M_i = \partial \{(x, s) \mid x \in \Sigma, \ s < u_i(x)\} \cap \left(\overline{B_i(p) \setminus K} \times \mathbb{R} \right).$$

We claim

$$S \cap M_i = \emptyset \qquad \text{for each } i \geqslant 1.$$

Since M_i converges to $\Sigma \times \{\tau\}$ as $i \to \infty$, $S \cap (\Sigma \times \{\tau\}) = \emptyset$ from the maximum principle. However, this contradicts to that $\tau \in (0, \delta)$ and $S \cap (\Sigma \times \{\delta'\}) \neq \emptyset$ for any $\delta' \in (0, \delta)$. This finishes the proof of Theorem 10.

A complete parabolic manifold must be M-parabolic, but a M-parabolic manifold may not be parabolic in general. However, they are equivalent in the following case.

Theorem 13 ([32]) *A M-parabolic manifold Σ is parabolic provided Σ satisfies the volume doubling property (3.1) and the (1,1)-Poincaré inequality (3.2) for any $x \in \Sigma$ and any $r \in (0, 1]$.*

Its proof relies on the convexity of volume functional of graphs, and a suitable average of functions. Let

$$f_\lambda(x) = \left(\int_0^\lambda \text{vol}(B_\tau(x)) d\tau \right)^{-1} \int_0^\lambda \left(\int_{B_\tau(x)} f \right) d\tau, \qquad x \in \Sigma.$$

Then we have

$$|Df_\lambda(x)| \leqslant c\lambda^{-1} \sup_{B_\lambda(x)} |f|,$$

$$\sqrt{1 + |Df_\lambda|^2(x)} - 1 \leqslant \frac{c}{\text{vol}(B_\lambda(x))} \int_{B_\lambda(x)} \left(\sqrt{1 + |Df|^2} - 1 \right).$$

(5.4)

Here, $c \geqslant 1$ is a general constant, which may change from line to line. For each $i \geqslant 2$, let u_i be a BV solution on Σ associated with $\text{cap}_1(\overline{B}_1(p), B_i(p))$. Let

$$u_i^*(x) = \left(\int_0^{1/2} \mathcal{H}^n(B_t(x)) dt \right)^{\frac{1}{2}} \int_0^{1/2} \left(\int_{B_t(x)} u_i \right) dt,$$

then with Fubini's theorem, we can deduce

$$c^{-1} \int_\Sigma |Du_i^*|^2 \leqslant \int_\Sigma \left(\sqrt{1 + |Du_i^*|^2} - 1 \right) \leqslant c \int_\Sigma \left(\sqrt{1 + |Du_i|^2} - 1 \right).$$

This means that M-parabolicity implies parabolicity.

Corollary 14 *A M-parabolic manifold of Ricci curvature bounded below is parabolic.*

Let Σ satisfy the volume doubling property (3.1) and the (1,1)-Poincaré inequality (3.2). Recalling (see Proposition 2 in [43] for instance) that the minimal positive Green function G satisfies

$$\inf_{x \in \partial B_r(p)} G(x,p) \leqslant \frac{1}{\text{cap}(\overline{B}_r(p))} \leqslant \sup_{x \in \partial B_r(p)} G(x,p).$$

(5.5)

Holopainen proved that the minimal positive Green function $G(x,p)$ on Σ satisfies

$$c^{-1} \int_{d(x,p)}^\infty \frac{r dr}{\text{vol}(B_r(p))} \leqslant G(x,p) \leqslant c \int_{d(x,p)}^\infty \frac{r dr}{\text{vol}(B_r(p))}.$$

(5.6)

which generalized Li-Yau's result [58] on manifolds of $Ric \geqslant 0$. Combining Harnack's inequality of harmonic functions and (5.5)(5.6), we can prove

Lemma 15 ([32]) *For any ball $B_r(p) \subset \Sigma$ and $r \geqslant t > 0$, there holds*

$$c^{-1} \int_r^\infty \frac{s\,ds}{\operatorname{vol}(B_s(p))} \leqslant \frac{t^2}{\operatorname{cap}_t(\overline{B}_r(p))} \leqslant c \int_r^\infty \frac{s\,ds}{\operatorname{vol}(B_s(p))}. \tag{5.7}$$

Combining (3.3) and Lemma 15, we can deduce the following asymptotic estimates.

Theorem 16 ([32]) *Given a compact set $K \subset \Sigma$ and $t > 0$, let u be a BV solution on Σ associated with $\operatorname{cap}_t(K)$. Then for any $r \geqslant \max\{t, 2\operatorname{diam}(K)\}$, $p \in K$ and $x \in \partial B_r(p)$ there holds*

$$\frac{\operatorname{cap}_t(K)}{ct} \int_r^\infty \frac{s\,ds}{\operatorname{vol}(B_s(p))} \leqslant u(x) \leqslant c\frac{\operatorname{cap}_t(K)}{t} \int_r^\infty \frac{s\,ds}{\operatorname{vol}(B_s(p))}.$$

Corollary 17 *If u is a minimal graphic function over $\Sigma \setminus B_1(p)$ with values in $(0,1]$ and $\lim\limits_{x \to \infty} u(x) = 0$. Denote $\lambda = \inf\limits_{\partial B_2(p)} u$. Then*

$$c^{-1}\lambda \int_{d(x,p)}^\infty \frac{s\,ds}{\operatorname{vol}(B_s(p))} \leqslant u(x) \leqslant c\lambda \int_{d(x,p)}^\infty \frac{s\,ds}{\operatorname{vol}(B_s(p))}. \tag{5.8}$$

6   The strategy of the proof of Theorem 4

The whole proof of Theorem 4 is composed of several arguments by contradiction combining Cheeger-Colding theory [11–14] and the current theory from geometric measure theory. The order of the proof of Theorem 4 is following:

1.   prove (4.5) for area-minimizing hypersurfaces in a special class of manifolds;

2.   prove (4.6) for a special class of area-minimizing hypersurfaces in a special class of manifolds using i);

3.   prove (4.6) for area-minimizing hypersurfaces in a special class of manifolds using i)ii);

4. prove (4.5) for minimal graphs over manifolds satisfying (4.4) using i)iii);

5.   prove (4.6) for minimal graphs over manifolds satisfying (4.4) using iii)iv).

More precisely, our proof of Theorem 4 is divided into three sections. In the first section, we prove the two inequalities (4.5)(4.6) for area-minimizing hypersurfaces in almost Euclidean spaces. Here, the spaces are complete manifolds of almost nonnegative Ricci curvature whose large balls are very close to Euclidean balls in the Gromov-Hausdorff sense. Then we argue by contradiction, and empoly modified C^1-mappings as Gromov-Hausdorff approximations essentially from Colding [17]. With the mappings, we are able to use the current theory from geometric measure theory, and get a Sobolev inequality on area-minimizing hypersurfaces in almost Euclidean spaces. Furthermore, through outward minimizing sets in area-minimizing hypersurfaces, using the mappings mentioned above and the current theory, we can deduce the Neumann-Poincaré inequality on area-minimizing hypersurfaces in almost Euclidean spaces based on our Sobolev inequality.

In the second section, we consider a sequence of minimal graphs M_i over $B_1(p_i)$ in $B_1(p_i) \times \mathbb{R}$, where $B_1(p_i)$ are geodesic balls with Ricci curvature uniformly bounded below. Suppose that $B_1(p_i)$ has a non-collapsing limit $B_1(p_\infty)$ with Gromov-Hausdorff approximations $\Psi_i : B_1(p_i) \to B_1(p_\infty)$, and $\Psi_i(M_i)$ converges to a limit M_∞ in $B_1(p_\infty) \times \mathbb{R}$ in the Hausdorff sense. We prove that any tangent cone of M_∞ is a metric cone in a tangent cone of $B_1(p_\infty) \times \mathbb{R}$, where we utilize the geometry of manifolds of Ricci curvature bounded below and the property of minimal graphs as well as some results in [30].

In the third section, we deal with the general Sobolev inequality on minimal graphs through an argument by contradiction with Cheeger-Colding theory. From [29], the cross section of every tangent cone of limits of minimal graphs is connected in the sense of splitting off a Euclidean factor isometrically. Combining tangent cones of M_∞ mentioned above and dimension estimate of singular sets of Ricci limit space, through dimension reduction argument, we are able to reduce the Sobolev inequality to the Neumann-Poincaré inequality on area-minimizing hypersurfaces in almost Euclidean spaces already established. In a similar manner, we can get the Neumann-Poincaré inequality on minimal graphs in manifolds with Ricci curvature bounded below.

7 Open problems

We have the following natural problems based on our previous results.

1. Whether Neumann-Poincaré inequality (4.6) holds without the condition $\mathrm{vol}(B_1(p)) \geqslant v$ in (4.4). Namely, is the constant in (4.6) independent of v?

2. Our gradient estimate (4.8) needs the condition: $\mathrm{vol}(B_r(x)) \geqslant vr^n$. Whether we can remove it? Namely, is the constant in (4.8) independent of v?

3. Whether M-parabolic condition in Theorem 10 is sharp? Namely, does the half-space property not hold for every M-nonparabolic manifold?

References

[1] F. J. Almgren Jr., *Some interior regularity theorems for minimal surfaces and an extension of Bernstein's theorem*, Ann. of Math. **85** (1966), 277–292.

[2] M. Anderson, L. Rodriguez, *Minimal surfaces and 3-manifolds with nonnegative Ricci curvature*, Math. Ann. **284**(1989), 461–475.

[3] M. T. Anderson, *The L^2 structure of moduli spaces of Einstein metrics on 4-manifolds*, Geom. Funct. Anal. **2** (1992), 29–89.

[4] E. Bombieri, *Theory of minimal surfaces and a counter-example to the Bernstein conjecture in high dimensions*, Notes of Lectures Held at the Courant Institute, New York University, 1970.

[5] E. Bombieri, E. De Giorgi and M. Miranda, *Una maggiorazione a priori relativa alle ipersuperfici minimali non parametriche*, Arch. Ration. Mech. Anal. **32** (1969), 255–267.

[6] E. Bombieri, E. Giusti, *Harnack's inequality for elliptic differential equations on minimal surfaces*, Invent. Math. **15** (1972), 24–46.

[7] E. Bombieri, E. De Giorgi and E. Giusti, *Minimal cones and the Bernstein problem*, Invent. Math. **7** (1969), 243–268.

[8] S. Brendle, *Sobolev inequalities in manifolds with nonnegative curvature*, Comm. Pure. Appl. Math. **76**(2023), no.9, 2192–2218.

[9] P. Buser, *A note on the isoperimetric constant*, Ann. Scient. Ec. Norm. Sup. **15** (1982), 213–230.

[10] J. B. Casteras, E. Heinonen, I. Holopainen, *Existence and non-existence of minimal graphic and p-harmonic functions*, Proc. Roy. Soc. Edinburgh Sect. A **150** (2020), no. 1, 341–366.

[11] J. Cheeger, T. H. Colding, *Lower bounds on Ricci curvature and the almost rigidity of warped products*, Ann. Math. **144** (1996), 189–237.

[12] J. Cheeger, T. H. Colding, *On the structure of spaces with Ricci curvature bounded below. I*, J. Differential Geom. **46** (1997), no. 3, 406–480.

[13] J. Cheeger, T. H. Colding, *On the structure of spaces with Ricci curvature bounded below. II*, J. Differential Geom. **54** (2000), no. 1, 15–35.

[14] J. Cheeger, T. H. Colding, *On the structure of spaces with Ricci curvature bounded below. III*, J. Differential Geom. **54** (2000), no. 1, 37–74.

[15] J. Cheeger, T. H. Colding, W. P. Minicozzi II, *Linear growth harmonic functions on complete manifolds with nonnegative Ricci curvature*, Geom. Funct. Anal. **5** (1995), no. 6, 948–954.

[16] J. Cheeger, A. Naber, *Quantitative stratification and the regularity of harmonic maps and minimal currents*, Comm. Pure Appl. Math. **66** (2013), 965–990.

[17] T. H. Colding, *Ricci curvature and volume convergence*, Ann. Math. **145** (1997), 477–501.

[18] T. H. Colding, *New monotonicity formulas for Ricci curvature and applications. I*, Acta Math., **209** (2012), 229–263.

[19] T. H. Colding and W. P. Minicozzi II, *Large scale behavior of kernels of Schrodinger operators*, Amer. J. Math. **119** (1997), no.6, 1355–1398.

[20] T. H. Colding and W. P. Minicozzi II, *Harmonic functions on manifolds*, Ann. Math.(2), **146**(1997), no.3, 725–747.

[21] G. Colombo, M. Magliaro, L. Mari, et al., *Bernstein and half-space properties for minimal graphs under Ricci lower bounds*, Inter. Math. Res. Not., **2022**(2022), no.23, 18256–18290.

[22] G. Colombo, E.S. Gama, L. Mari, et al., *Non-negative Ricci curvature and Minimal graphs with linear growth*, arXiv:2112.09886.

[23] C. Croke, *Some isoperimetric inequalities and eigenvalue estimates*, Ann. Sci. École. Norm. Sup. **13** (1980), no.4, 419–435.

[24] E. De Giorgi, *Frontiere orientate di misura minima*, Sem. Mat. Scuola Norm. Sup. Pisa. 1960/1961.

[25] E. De Giorgi, *Una estensione del teorema di Bernstein*, Ann. Sci. Norm. Sup. Pisa **19** (1965), 79–85.

[26] Q. Ding, J.Jost and Y.L.Xin, *Minimal graphic functions on manifolds of non-negative Ricci curvature*, Comm. Pure Appl. Math. **69**(2016), no.2, 323–371.

[27] Q. Ding, J.Jost and Y.L.Xin, *Existence and non-existence of area-minimizing hypersurfaces in manifolds of non-negative Ricci curvature*, Amer. J. Math. **138** (2016), no.2, 287–327.

[28] Q. Ding, *Liouville-type theorems for minimal graphs over manifolds*, Analysis & PDE **14**(2021), no.6, 1925–1949.

[29] Q. Ding, *Minimal hypersurfaces in manifolds of Ricci curvature bounded below*, J. Reine. Angew. Math. **791**(2022), 247–282.

[30] Q. Ding, *Area-minimizing hypersurfaces in manifolds of Ricci curvature bounded below*, J. Reine. Angew. Math. **798**(2023), 193–236.

[31] Q. Ding, *Poincaré inequality on minimal graphs over manifolds and applications*, arXiv:2111.04458.

[32] Q. Ding, *Capacity for minimal graphs over manifolds and the half-space property*, arXiv:2306.15137.

[33] J. M. Espinar and H. Rosenberg, *Frankel property and maximum principle at infinity for complete minimal hypersurfaces*, arXiv: 2211.06392.

[34] H. Federer, *Geometric Measure Theory*, Berlin-Heidelberg-New York: Springer-Verlag, 1969.

[35] H. Federer, W.H. Fleming, *Normal and integral currents*, Ann. Math. **72** (1960), no.2, 458–520.

[36] H. Federer, *The singular sets of area minimizing rectifiable currents with codimension one and of area minimizing flat chains modulo two with arbitrary codimension*, Bull. Amer. Math. Soc. **76** (1970), 767–771.

[37] R. Finn, *On equations of minimal surface type*, Ann. Math., **60**(1954), no.2, 397–416.

[38] R. Finn, *New estimates for equations of minimal surface type*, Arch. Rational Mech. Anal. **14**(1963), 337–375.

[39] D. Fischer-Colbrie, R. Schoen, *The structure of complete stable minimal surfaces in 3-manifolds of nonnegative scalar curvature*, Comm. Pure Appl. Math. **33** (1980), 199–211.

[40] W. Fleming, *On the oriented Plateau problem*, Rend Circ. Mat. Palermo **11** (1962), 1–22.

[41] T. Frankel, *On the fundamental group of a compact minimal submanifold*, Ann. Math. **83** (1966), 68–73.

[42] E. Giusti, *Minimal Surfaces and Functions of Bounded Variation*, Boston: Birkhäuser, 1984.

[43] A. A. Grigor'yan, *On the existence of positive fundamental solution of the Laplace equation on Riemannian manifolds*, (in Russian) Matem. Sbornik, **128** (1985) no.3, 354-363. Engl. transl. Math. USSR Sb., **56** (1987) 349–358.

[44] A. A. Grigor'yan, *Analytic and geometric background of recurrence and non-explosion of the Brownian motion on Riemannian manifolds*, Bull. Amer. Math. Soc. (N.S.) **36** (1999), no. 2, 135–249.

[45] A. Grigor'yan, *Estimates of heat kernels on Riemannian manifolds*//E. B. Davies and Y. Safalov, Spectral Theory and Geometry (Edinburgh, 1998), London Math. Soc. Lecture Note Ser., 273, Cambridge: Cambridge University Press, 1999, 140–225.

[46] E. Heintze, H. Karcher, *A general comparison theorem with applications to volume estimates for submanifolds*, Ann. Sci. Ecole Norm. Sup. **11** (1978), no. 4, 451–470.

[47] D. Hoffman and W. H. Meeks, III, *The strong halfspace theorem for minimal surfaces*, Invent. Math. **101** (1990), no. 2, 373–377.

[48] D. Hoffman and J. Spruck, *Sobolev and isoperimetric inequalities for Riemannian submanifolds*, Comm. Pure Appl. Math. **27** (1974), 715–727.

[49] J. C. C. Hoyos, *Poincaré and Sobolev type inequalities for intrinsic rectifiable varifolds*, arXiv:2001.09256.

[50] Y. Itokawa and R. Kobayashi, *Minimizing currents in open manifolds and the $n - 1$ homology of nonnegatively Ricci curved manifolds*, Amer. J. Math. **121** (1999), no. 6, 1253–1278.

[51] H. Jenkins and J. Serrin, *The Dirichlet problem for the minimal surface equation in higher dimensions*, J. Reine Angew. Math. **229** (1968), 170–187.

[52] N. Korevaar, *An easy proof of the interior gradient bound for solutions of the precribed mean curvature equation.* Proc. Symp. Pure Math. **45**, AMS (1986), 81–89.

[53] R. Langevin, H. Rosenberg, *A maximum principle at infinity for minimal surfaces and applications*, Duke Math. J. **57** (1988) no. 3, 819–828.

[54] P. Li, *Harmonic sections of polynomial growth*, Math. Res. Lett. **4** (1997), no.1, 35–44.

[55] P. Li, *Harmonic Functions and Applications to Complete Manifolds*, University of California, Irvine, 2004, preprint.

[56] P. Li, L. F. Tam, and J. P. Wang, *Sharp bounds for the Green's function and the heat kernel*, Math. Res. Lett. **4**(1997), 589–602.

[57] P. Li, J. P. Wang, *Mean value inequalities*, Indiana Univ. Math. J. **48** (1999), no. 4, 1257–1283.

[58] P. Li and S. T. Yau, *On the parabolic kernel of the Schrödinger operator*, Acta Math. **156** (1986), 153–201.

[59] F. H. Lin, *Regularity for a Class of Parametric Obstacle Problems*, Ph.D. thesis, University of Minnesota, July 1985.

[60] F. H. Lin, X. P. Yang, *Geometric Measure Theory: An Introduction*, Beijing: Science Press/New York: International Press, Boston, 2002.

[61] G. Liu, *3-manifolds with nonnegative Ricci curvature*, Invent. Math. **193** (2013), no. 2, 367–375.

[62] W. H. Meeks Ⅲ and H. Rosenberg, *Maximum principles at infinity*, J. Diff. Geom. **79** (2008) no. 1, 141–165.

[63] J. Michael and L. M. Simon, *Sobolev and mean-value inequalities on generalized submanifolds of $\mathbb{R}^n$*, Comm. Pure Appl. Math., **26** (1973), 361–379.

[64] M. Miranda, *Disuguaglianze di Sobolev sulle ipersuperfici minimali*, Rend. Sem. Mat. Univ. Padova, **38**, 1967.

[65] A. Naber and D. Valtorta, *The singular structure and regularity of stationary varifolds*, J. Eur. Math. Soc. **22** (2020), no. 10, 3305–3382.

[66] E.R. Reifenberg, *Solution of the Plateau problem for m-dimensional surfaces of varying topological type*, Acta Math. **104** (1960), 1–92.

[67] H. Rosenberg, F. Schulze and J. Spruck, *The half-space property and entire positive minimal graphs in $M \times \mathbb{R}$*, J. Diff. Geom. **95** (2013), 321–336.

[68] C. Scharrer, *Some geometric inequalities for varifolds on Riemannian manifolds based on monotonicity identities*, Ann. Glob. Anal. Geom. **61**(2022), 691–719.

[69] R. Schoen, L. Simon, and S. T. Yau, *Curvature estimates for minimal hypersurfaces*, Acta Math. **134** (1975), 275–288.

[70] R. Schoen and S. T. Yau, *On the proof of the positive mass conjecture in general relativity*, Comm. math. Phys. **65** (1979), 45–76.

[71] R. Schoen and S. T. Yau, *Proof of the positive mass theorem. II*, Comm. Math. Phys. **79** (1981), no. 2, 231–260.

[72] L. Simon, Thesis, University of Adelaide, 1971.

[73] L. Simon, *Lectures on Geometric Measure Theory*, Proceedings of the Center for Mathematical Analysis, Australian National University, Vol. 3, 1983.

[74] L. Simon, *Entire solutions of the minimal surface equation*, J. Differential Geom. **30** (1989), 643–688.

[75] L. Simon, *Uniqueness of some cylindrical tangent cones*, Comm. Anal. Geom. **2** (1994), no.1, 1–33.

[76] J. Simons, *Minimal varieties in Riemannian manifolds*, Ann. Math. **88** (1968), 62–105.

[77] J. Spruck, *Interior gradient estimates and existence theorems for constant mean curvature graphs in $M^n \times \mathbb{R}$*, Pure Appl. Math. Q. **3** (2007), no. 3, Special Issue: In honor of Leon Simon. Part 2, 785–800.

[78] G. Székelyhidi, *Uniqueness of certain cylindrical tangent cones*, arXiv:2012.02065.

[79] X. J. Wang, *Interior gradient estimates for mean curvature equations*, Math. Z., **228** (1998), 73–81.

[80] S. T. Yau, *Harmonic functions on complete manifolds*, Comm. Pure Appl. Math. **28**(1975), 201–228.

Deformation Quantization and Noncommutative Differential Geometry

HAOYUAN GAO

Shanghai Center for Mathematical Sciences, Fudan University, Shanghai, China

E-mail: hy_gao@fudan.edu.cn

XIAO ZHANG

Institute of Mathematics, Academy of Mathematics and Systems Sciences, Chinese Academy of Sciences, Beijing, China

Guangxi Center for Mathematical Research, Guangxi University, Nanning, China

E-mail: xzhang@amss.ac.cn; xzhang@gxu.edu.cn

Abstract This is a survey of the noncommutative differential geometry in the framework of deformation quantization and its application in the study of quantum gravity.

1 Introduction

Gravity is essentially a theory of spacetime geometry. In the concept of quantum effects of gravity, the Heisenberg uncertainty relations would result in noncommutativity of spacetime variables for sufficiently small distances. In [17, 21] the first attempts to quantize spacetimes were made, which are referred to as Snyder's quantum space times and Yang's quantum phase spaces [13, 14]. In their approach, spacetime variables were represented by Hermitian operators with discrete eigenvalues. Alternatively, deformation quantization was developed, which deforms the commutative algebra of functions based on the pointwise product to noncommutative algebras of formal functions based on certain noncommutative products such as the Moyal product, but still keeps spacetime variables usual functions, cf. [3, 4].

In recent years, there have been intensive research activities on noncommutative gravity in the framework of deformation quantization, cf. [1, 2, 16] and references therein, where general relativity is adopted to the noncommutative setting in an intuitive way, as pointed out in [18].

In [5, 19, 20, 22], a mathematically rigorous and complete theory of noncommutative differential geometry was developed on a coordinate chart U of a (pseudo-) Riemannian manifold. The idea is to embed U isometrically into a flat (pseudo-) Euclidean space and use the isometric embedding to construct the noncommutative analogues of metric, connection and curvature. They yield the noncommutative Einstein field equations. It was found that the deformation quantization of the Schwarzschild metric does not depend on time and yields an unevaporated quantum black hole [19], and the quantum fluctuation of the plane-fronted gravitational wave is the exact solution of the noncommutative vacuum Einstein field equations [20].

In [8, 23] an intrinsic formulation of the noncommutative differential geometry developed in [5, 19, 20, 22] was provided. This intrinsic formulation yields geometric definitions of covariant derivatives of noncommutative metrics and curvatures, as well as the noncommutative version of Bianchi identities. Moreover, in [8] it was shown that if a noncommutative metric and chiral coefficients satisfy certain conditions which hold automatically for quantum fluctuation given by an isometric embedding, then the two noncommutative Ricci curvatures are essentially equivalent. The renormalization of quantization of spherically symmetric spacetimes was also discussed in [8], generalizing some results in [19]. The study of noncommutative differential geometry in the framework of general star products began in [8].

2 Moyal product and noncommutative differential geometry

Let M be an n-dimensional differentiable manifold and $U \subset M$ be a coordinate chart equipped with coordinates $(x^1, \cdots, x^n)$. Let $\hbar$ be the Planck constant viewed as an indeterminate. Denote $\mathbb{R}[[\hbar]]$ the ring of formal power series in $\hbar$ with real coefficients, and $\mathcal{A}_U$ the set of formal power series in $\hbar$ with coefficients being real smooth functions on U

$$\mathcal{A}_U = C^\infty(U)[[\hbar]] = \Big\{ \sum_{q=0}^{\infty} f_q \hbar^q \Big| f_q \in C^\infty(U) \Big\}.$$

$\mathcal{A}_U$ is an $\mathbb{R}[[\hbar]]$-module.

Throughout the paper, all the indices i, j, k, l, $\cdots$, range from 1 to n, $q \in \mathbb{N}_0$. We also use the Einstein summation convention. For an element $f \in \mathcal{A}_U$, we denote $f[q] \in C^\infty(U)$ the coefficient of the q-th order item of f, i.e.

$$f = \sum_{q=0}^{\infty} f[q] \hbar^q.$$

Let (θ^{ij}) be a skew-symmetric $n \times n$ real matrix. Given two smooth functions u, v on U, the Moyal product of u and v with respect to (θ^{ij}) is defined as

$$(u * v)(x) = \Big[\exp(\hbar \theta^{ij} \partial_i \partial_j') u(x) v(x') \Big]_{x=x'},$$

where x and x' denote the same coordinate system and $\partial_i = \dfrac{\partial}{\partial x^i}$, $\partial_i' = \dfrac{\partial}{\partial (x')^i}$. It is clearly by definition that

$$u * v \in \mathcal{A}_U.$$

Extending by $\mathbb{R}[[\hbar]]$-bilinearity, the Moyal product provides an associative $\mathbb{R}[[\hbar]]$-bilinear product on $\mathcal{A}_U$, cf. [12]. $\mathcal{A}_U$ equipped with the Moyal product is called the Moyal algebra on U. It is a formal deformation of the algebra of real smooth functions on U [9]. Moreover, it is a deformation quantization of the constant Poisson structure $\dfrac{1}{2}\theta^{ij} \partial_i \wedge \partial_j$ [4,11].

Extend ∂_i to $\mathcal{A}_U$ by $\mathbb{R}[[\hbar]]$-linearity, the Moyal product satisfies

(i) Noncommutativity: $[x^i, x^j] = x^i * x^j - x^j * x^i = 2\hbar \theta^{ij}$;

(ii) Leibniz rule: $\partial_i(u * v) = (\partial_i u) * v + u * (\partial_i v)$, for $u, v \in \mathcal{A}_U$.

Denote $E_i = \tilde{E}_i = \partial_i$, $1 \leqslant i \leqslant n$. The noncommutative left (resp. right) tangent bundle $\mathcal{T}_U$ (resp. $\tilde{\mathcal{T}}_U$) on U is the free left (resp. right) $\mathcal{A}_U$-module with basis $\{E_1, \cdots, E_n\}$ (resp. $\{\tilde{E}_1, \cdots, \tilde{E}_n\}$), i.e.,

$$\mathcal{T}_U = \Big\{ a^i * E_i \Big| a^i \in \mathcal{A}_U, \ a^i * E_i = 0 \Longleftrightarrow a^i = 0. \Big\},$$

$$\tilde{\mathcal{T}}_U = \Big\{ \tilde{E}_i * a^i \Big| a^i \in \mathcal{A}_U, \ \tilde{E}_i * a^i = 0 \Longleftrightarrow a^i = 0. \Big\}.$$

An element of $\mathcal{T}_U$ (resp. $\tilde{\mathcal{T}}_U$) is called a left (resp. right) vector field.

A noncommutative metric g on U is a homomorphism of two-sided $\mathcal{A}_U$-modules

$$g : \mathcal{T}_U \otimes_{\mathbb{R}[[\hbar]]} \tilde{\mathcal{T}}_U \longrightarrow \mathcal{A}_U,$$

such that the matrix

$$(g_{ij}) \in \mathcal{A}_U^{n \times n}, \quad g_{ij} = g(E_i, \tilde{E}_j)$$

is invertible, i.e., there exists a unique matrix $(g^{ij}) \in \mathcal{A}_U^{n \times n}$ such that

$$g_{ik} * g^{kj} = g^{jk} * g_{ki} = \delta_i^j.$$

In [5, 8], it was shown that (g_{ij}) is invertible if and only if the real matrix $(g_{ij}[0](x))$ is invertible for any $x \in U$.

The noncommutative metric g induces the dual basis $\tilde{E}^i$ (resp. E^i) of E_j (resp. $\tilde{E}_j$) in the sense that

$$g(E^i, \tilde{E}_j) = g(E_j, \tilde{E}^i) = \delta_j^i.$$

Then the noncommutative left and right cotangent bundles are defined as free $\mathcal{A}_U$-modules

$$\mathcal{T}_U^* = \left\{ a_i * E^i \,\middle|\, a_i \in \mathcal{A}_U, \, a_i * E^i = 0 \Longleftrightarrow a_i = 0. \right\},$$

$$\tilde{\mathcal{T}}_U^* = \left\{ \tilde{E}^i * a_i \,\middle|\, a_i \in \mathcal{A}_U, \, \tilde{E}^i * a_i = 0 \Longleftrightarrow a_i = 0. \right\}.$$

The noncommutative metric g acts as an element of $\tilde{\mathcal{T}}_U^* \otimes_{\mathcal{A}_U} \mathcal{T}_U^*$,

$$g = \tilde{E}^i \otimes g_{ij} * E^j = \tilde{E}^i * g_{ij} \otimes E^j. \tag{2.1}$$

The inverse matrix (g^{ij}) can be viewed as a homomorphism of two-sided modules

$$g^{-1} : \mathcal{T}_U^* \otimes_{\mathbb{R}[[\hbar]]} \tilde{\mathcal{T}}_U^* \longrightarrow \mathcal{A}_U$$

such that

$$g^{-1}(E^i, \tilde{E}^j) = g^{ij}.$$

Similarly, g^{-1} acts as an element of $\tilde{\mathcal{T}}_U \otimes_{\mathcal{A}_U} \mathcal{T}_U$,

$$g^{-1} = \tilde{E}_i \otimes g^{ij} * E_j = \tilde{E}_i * g^{ij} \otimes E_j. \tag{2.2}$$

A noncommutative left (resp. right) connection ∇ is a map

$$\nabla : \mathcal{T}_U \longrightarrow \tilde{\mathcal{T}}_U^* \otimes_{\mathcal{A}_U} \mathcal{T}_U \quad (\text{resp. } \tilde{\nabla} : \tilde{\mathcal{T}}_U \longrightarrow \tilde{\mathcal{T}}_U \otimes_{\mathcal{A}_U} \mathcal{T}_U^*)$$

such that noncommutative left (resp. right) covariant derivatives

$$\nabla_i : \mathcal{T}_U \longrightarrow \mathcal{T}_U, \quad (\text{resp. } \tilde{\nabla}_i : \tilde{\mathcal{T}}_U \longrightarrow \tilde{\mathcal{T}}_U);$$

defined by

$$\nabla_i V = g(E_i, \tilde{E}^k) * W_k \quad (\text{resp. } \tilde{\nabla}_i \tilde{V} = \tilde{W}_k * g(E^k, \tilde{E}_i))$$

for any

$$\nabla V = \tilde{E}^k \otimes W_k, \ W_k \in \mathcal{T}_U \quad (\text{resp. } \tilde{\nabla}\tilde{V} = \tilde{W}_k \otimes E^k, \ \tilde{W}_k \in \tilde{\mathcal{T}}_U)$$

satisfy

(i) $\mathbb{R}[[\hbar]]$-linearity: For $a, b \in \mathbb{R}[[\hbar]]$, $V, W \in \mathcal{T}_U$ (resp. $\tilde{V}, \tilde{W} \in \tilde{\mathcal{T}}_U$),

$$\nabla_i(aV + bW) = a\nabla_i V + b\nabla_i W,$$

$$\tilde{\nabla}_i(a\tilde{V} + b\tilde{W}) = a\tilde{\nabla}_i \tilde{V} + b\tilde{\nabla}_i \tilde{W};$$

(ii) Leibniz rule: For $f \in \mathcal{A}_U$, $V \in \mathcal{T}_U$ (resp. $\tilde{V} \in \tilde{\mathcal{T}}_U$),

$$\nabla_i(f * V) = (\partial_i f) * V + f * \nabla_i V,$$

$$\tilde{\nabla}_i(\tilde{V} * f) = \tilde{V} * (\partial_i f) + \tilde{\nabla}_i \tilde{V} * f.$$

The left and right connections are uniquely determined by connection coefficients Γ_{ij}^k and $\tilde{\Gamma}_{ij}^k$, which are elements of $\mathcal{A}_U$

$$\nabla_i E_j = \Gamma_{ij}^k * E_k, \quad \tilde{\nabla}_i \tilde{E}_j = \tilde{E}_k * \tilde{\Gamma}_{ij}^k.$$

A noncommutative connection consists of a noncommutative left connection ∇ and a noncommutative right connection $\tilde{\nabla}$. It induces a connection on the noncommutative cotangent bundles

$$\nabla_i : \mathcal{T}_U^* \longrightarrow \mathcal{T}_U^*, \quad \nabla_i E^j = \Gamma_{ik}^{*j} * E^k,$$

$$\tilde{\nabla}_i : \tilde{\mathcal{T}}_U^* \longrightarrow \tilde{\mathcal{T}}_U^*, \quad \tilde{\nabla}_i \tilde{E}^j = \tilde{E}^k * \tilde{\Gamma}_{ik}^{*j},$$

satisfying the compatible conditions

$$\partial_k g(E_i, \tilde{E}^j) = g(\nabla_k E_i, \tilde{E}^j) + g(E_i, \tilde{\nabla}_k \tilde{E}^j),$$

$$\partial_k g(E^i, \tilde{E}_j) = g(\nabla_k E^i, \tilde{E}_j) + g(E^i, \tilde{\nabla}_k \tilde{E}_j).$$

Thus

$$\nabla_i E^j = -\tilde{\Gamma}^j_{ik} * E^k, \quad \tilde{\nabla}_i \tilde{E}^j = -\tilde{E}^k * \Gamma^j_{ik}.$$

Denote

$$\Gamma_{ijk} = \Gamma^l_{ij} * g_{lk}, \quad \tilde{\Gamma}_{ijk} = g_{kl} * \tilde{\Gamma}^l_{ij}.$$

Given a noncommutative metric g and a set of elements Υ_{ijk} of $\mathcal{A}_U$ with

$$\Upsilon_{ijk} = \Upsilon_{jik},$$

which are referred as the chiral coefficients. A noncommutative connection $\{\nabla, \tilde{\nabla}\}$ is canonical with respect to g and Υ_{ijk} [8, 23] if it satisfies

(i) Compatibility: $\partial_k g_{ij} = g(\nabla_k E_i, \tilde{E}_j) + g(E_i, \tilde{\nabla}_k \tilde{E}_j) = \Gamma_{kij} + \tilde{\Gamma}_{kji}$;

(ii) Torsion free: $\nabla_i E_j = \nabla_j E_i, \tilde{\nabla}_i \tilde{E}_j = \tilde{\nabla}_j \tilde{E}_i$;

(iii) Chirality: $\Gamma_{ijk} - \tilde{\Gamma}_{ijk} = \Upsilon_{ijk}$.

The torsion free condition implies

$$\Gamma^k_{ij} = \Gamma^k_{ji}, \quad \tilde{\Gamma}^k_{ij} = \tilde{\Gamma}^k_{ji}.$$

It is straightforward that [8]

$$2\Gamma_{ijk} = \partial_i g_{jk} + \partial_j g_{ki} - \partial_k g_{ij} + \Upsilon_{ikj} + \Upsilon_{jik} - \Upsilon_{kji}$$

$$= \partial_i g_{jk} + \partial_j g_{ki} - \partial_k g_{ji} + \Upsilon_{ijk}$$

$$= \partial_i \left(\frac{g_{jk} + g_{kj}}{2}\right) + \partial_j \left(\frac{g_{ki} + g_{ik}}{2}\right) - \partial_k \left(\frac{g_{ij} + g_{ji}}{2}\right) + \Upsilon_{ijk},$$

and

$$2\tilde{\Gamma}_{ijk} = \partial_i g_{jk} + \partial_j g_{ki} - \partial_k g_{ij} + \Upsilon_{ikj} - \Upsilon_{jik} - \Upsilon_{kji}$$

$$= \partial_i g_{jk} + \partial_j g_{ki} - \partial_k g_{ji} - \Upsilon_{ijk}$$

$$= \partial_i \left(\frac{g_{jk} + g_{kj}}{2}\right) + \partial_j \left(\frac{g_{ki} + g_{ik}}{2}\right) - \partial_k \left(\frac{g_{ij} + g_{ji}}{2}\right) - \Upsilon_{ijk}.$$

The left curvature operators $\mathcal{R}_{E_i E_j}$ and the right curvature operators $\tilde{\mathcal{R}}_{\tilde{E}_i \tilde{E}_j}$ can be defined as the following $\mathcal{A}_U$-linear operators [5]

$$\mathcal{R}_{E_i E_j} = [\nabla_i, \nabla_j] : \mathcal{T}_U \longrightarrow \mathcal{T}_U,$$

$$\tilde{\mathcal{R}}_{\tilde{E}_i \tilde{E}_j} = [\tilde{\nabla}_i, \tilde{\nabla}_j] : \tilde{\mathcal{T}}_U \longrightarrow \tilde{\mathcal{T}}_U.$$

For a canonical connection, the left Riemannian curvatures R_{lkij} and right Riemannian curvatures $\tilde{R}_{lkij}$ are defined as

$$R_{lkij} = g(\mathcal{R}_{E_i E_j} E_k, \tilde{E}_l), \quad \tilde{R}_{lkij} = -g(E_k, \tilde{\mathcal{R}}_{\tilde{E}_i \tilde{E}_j} \tilde{E}_l).$$

They satisfy [5]

$$R_{lkij} = -R_{lkji} = \tilde{R}_{lkij}, \quad R_{lkij} \not\equiv -R_{klij}.$$

Therefore the left curvatures are sufficient for the purpose. There are two Ricci curvatures R_{kj} and Θ_{il} obtained by contracting l, i and k, j in R_{lkij} respectively

$$R_{kj} = g(\mathcal{R}_{E_i E_j} E_k, \tilde{E}_l) * g^{li} = R_{lkij} * g^{li},$$

$$\Theta_{il} = g^{jk} * g(\mathcal{R}_{E_i E_j} E_k, \tilde{E}_l) = g^{jk} * R_{lkij}.$$

Raising the index at k and l respectively, we have Ricci curvatures

$$R_j^p = g^{pk} * g(\mathcal{R}_{E_i E_j} E_k, \tilde{E}_l) * g^{li} = g^{pk} * R_{lkij} * g^{li},$$

$$\Theta_i^p = g^{jk} * g(\mathcal{R}_{E_i E_j} E_k, \tilde{E}_l) * g^{lp} = g^{jk} * R_{lkij} * g^{lp}.$$

The two Ricci curvatures R_i^p and Θ_i^p are not equal to each other in the noncommutative case. But their traces coincide and yield the same scalar curvature

$$R = R_j^j = \Theta_i^i.$$

3   Covariant derivatives and noncommutative Bianchi identities

As in classical differential geometry, covariant derivatives of noncommutative metrics and curvatures also make sense. Using the tensor product expressions (2.1), (2.2), the covariant derivatives of a noncommutative metric g and its inverse g^{-1}, denoted by

$$\nabla_k g = \tilde{E}^i \otimes \nabla_k g_{ij} * E^j = \tilde{E}^i * \nabla_k g_{ij} \otimes E^j,$$

$$\nabla_k g^{-1} = \tilde{E}_i \otimes \nabla_k g^{ij} * E_j = \tilde{E}_i * \nabla_k g^{ij} \otimes E_j,$$

are defined as

$$\nabla_k g = \tilde{\nabla}_k \tilde{E}^i \otimes g_{ij} * E^j + \tilde{E}^i \otimes \nabla_k(g_{ij} * E^j)$$

$$= \tilde{\nabla}_k(\tilde{E}^i * g_{ij}) \otimes E^j + \tilde{E}^i * g_{ij} \otimes \nabla_k E^j,$$

$$\nabla_k g^{-1} = \tilde{\nabla}_k \tilde{E}_i \otimes g^{ij} * E_j + \tilde{E}_i \otimes \nabla_k(g^{ij} * E_j)$$

$$= \tilde{\nabla}_k(\tilde{E}_i * g^{ij}) \otimes E_j + \tilde{E}_i * g^{ij} \otimes \nabla_k E_j.$$

In [8], it was shown that if the connection is compatible with the metric g, then

$$\nabla_k g = 0, \quad \nabla_k g^{-1} = 0.$$

The noncommutative left (resp. right) covariant derivative along a left (resp. right) vector field $V = a^i * E_i$ (resp. $W = \tilde{E}^i * a^i$) with $a^i \in \mathcal{A}_U$ is defined as the $\mathbb{R}[[\hbar]]$-linear map

$$\nabla_V : \mathcal{T}_U \to \mathcal{T}_U \quad (\text{resp. } \tilde{\nabla}_W : \tilde{\mathcal{T}}_U \to \tilde{\mathcal{T}}_U)$$

given by

$$\nabla_V X = a^i * (\nabla_i X) \quad (\text{resp. } \tilde{\nabla}_W Y = (\tilde{\nabla}_i Y) * a^i)$$

for $X \in \mathcal{T}_U$ (resp. $Y \in \tilde{\mathcal{T}}_U$).

For left and right tangent vectors

$$V = v^i * E_i, \quad W = w^j * E_j, \quad \tilde{V} = \tilde{E}_i * \tilde{v}^i, \quad \tilde{W} = \tilde{E}_j * \tilde{w}^j,$$

the noncommutative left and right curvature operators for left and right tangent vectors can be formally defined as

$$\mathcal{R}_{VW} = [\nabla_V, \nabla_W] - \nabla_{[V,W]},$$

$$\tilde{\mathcal{R}}_{\tilde{V}\tilde{W}} = [\tilde{\nabla}_{\tilde{V}}, \tilde{\nabla}_{\tilde{W}}] - \tilde{\nabla}_{[\tilde{V},\tilde{W}]}.$$

But the Lie brackets $[V, W], [\tilde{V}, \tilde{W}]$ do not give rise to noncommutative vector fields unless

$$[v^i, w^j] = [\tilde{v}^i, \tilde{w}^j] = 0.$$

Thus, if $V = E_i$ (resp. $\tilde{V} = \tilde{E}_i$) or $W = E_j$ (resp. $\tilde{W} = \tilde{E}_j$), then $\mathcal{R}_{VW} E_k \in \mathcal{T}_U$ (resp. $\tilde{\mathcal{R}}_{\tilde{V}\tilde{W}} \tilde{E}_k \in \tilde{\mathcal{T}}_U$) is well-defined. This suggests the definitions of noncommutative covariant derivatives of curvatures as follows.

Definition 3.1 ([8])   The covariant derivatives of noncommutative curvature operators are defined as follows.

$$
\begin{aligned}
(\nabla_k \mathcal{R})_{E_i E_j} E_p =& \nabla_k (\mathcal{R}_{E_i E_j} E_p) - \mathcal{R}_{(\nabla_{E_k} E_i) E_j} E_p \\
& - \mathcal{R}_{E_i (\nabla_{E_k} E_j)} E_p - \mathcal{R}_{E_i E_j} (\nabla_{E_k} E_p).
\end{aligned}
$$

Definition 3.2 ([8])   The covariant derivatives of noncommutative curvature tensors, noncommutative Ricci curvatures and noncommutative scalar curvature are defined as follows.

$$
\begin{aligned}
\nabla_s R_{lkij} =& g\big((\nabla_s \mathcal{R})_{E_i E_j} E_k, \tilde{E}_l\big), \\
\nabla_s R_j^p =& g^{pk} * g\big((\nabla_s \mathcal{R})_{E_i E_j} E_k, \tilde{E}_l\big) * g^{li} = g^{pk} * \nabla_s R_{lkij} * g^{li}, \\
\nabla_s \Theta_i^p =& g^{jk} * g\big((\nabla_s \mathcal{R})_{E_i E_j} E_k, \tilde{E}_l\big) * g^{lp} = g^{jk} * \nabla_s R_{lkij} * g^{lp}, \\
\nabla_s R =& g^{jk} * g\big((\nabla_s \mathcal{R})_{E_i E_j} E_k, \tilde{E}_l\big) * g^{li}.
\end{aligned}
$$

The noncommutative covariant derivatives of right curvatures are defined in the same way.

For a canonical connection, we have the first (algebraic) Bianchi identity [8]

$$\mathcal{R}_{E_i E_j} E_k + \mathcal{R}_{E_j E_k} E_i + \mathcal{R}_{E_k E_i} E_j = 0,$$

and the second (differential) Bianchi identity [8]

$$(\nabla_i \mathcal{R})_{E_j E_k} E_p + (\nabla_j \mathcal{R})_{E_k E_i} E_p + (\nabla_k \mathcal{R})_{E_i E_j} E_p = 0.$$

The noncommutative Bianchi identities also hold for right curvatures. The second Bianchi identity implies that [8]

$$\nabla_i R_j^i + \nabla_i \Theta_j^i - \delta_j^i \nabla_i R = 0. \tag{3.1}$$

4　Equivalence of Ricci curvatures and Einstein field equations

The two noncommutative Ricci curvatures can be quite different in general, but in certain conditions, they are essentially equivalent. In [8] we gave a sufficient condition under which the two Ricci curvatures are equivalent and this condition holds automatically for quantum fluctuation given by an isometric embedding. More precisely, we have the following theorem.

Theorem 4.1 ([8])　*Let M be an n-dimensional smooth manifold and $U \subset M$ a coordinate chart. Let ∇, $\tilde{\nabla}$ be the canonical connection with respect to noncommutative metric g and chiral coefficients Υ_{ijk} on U. If g and Υ_{ijk} satisfy*

$$g_{ij}[2q] = g_{ji}[2q], \quad g_{ij}[2q+1] = -g_{ji}[2q+1], \tag{4.1}$$

and

$$\Upsilon_{ijk}[2q] = 0, \tag{4.2}$$

then the two Ricci curvatures are equivalent in the sense that

$$R_{ij}[2q] = \Theta_{ji}[2q], \quad R_{ij}[2q+1] = -\Theta_{ji}[2q+1], \tag{4.3}$$

and

$$R^i_j[2q] = \Theta^i_j[2q], \quad R^i_j[2q+1] = -\Theta^i_j[2q+1]. \tag{4.4}$$

Given a pseudo-Riemannian metric $g[0]$ in U, recall that $(U, g[0])$ can always be isometrically embedded into a pseudo-Euclidean space, cf. [15], i.e., there exists a differentiable map

$$X : U \longrightarrow \mathbb{R}^{p,m-p}$$

such that

$$g_{ij}[0] = \sum_{\alpha=1}^m \eta_{\alpha\alpha} \partial_i X^\alpha \cdot \partial_j X^\alpha,$$

where $\eta = \operatorname{diag}(-1, \cdots, -1, 1, \cdots, 1)$ is the flat metrics of $\mathbb{R}^{p,m-p}$ and X^α are components of X. The quantum fluctuation of $g[0]$ given by X is [5,8,23]

$$g\big(E_i, \tilde{E}_j\big) = \sum_{\alpha=1}^m \eta_{\alpha\alpha} \partial_i X^\alpha * \partial_j X^\alpha, \tag{4.5}$$

where $E_i = \tilde{E}_i = \partial_i$. It yields a canonical connection with the chiral and connection coefficients

$$\Upsilon_{ijk} = \sum_{\alpha=1}^{m} \eta_{\alpha\alpha} \big(\partial_i \partial_j X^\alpha * \partial_k X^\alpha - \partial_k X^\alpha * \partial_i \partial_j X^\alpha \big), \tag{4.6}$$

$$\Gamma_{ijk} = \sum_{\alpha=1}^{m} \eta_{\alpha\alpha} \partial_i \partial_j X^\alpha * \partial_k X^\alpha, \tag{4.7}$$

$$\tilde{\Gamma}_{ijk} = \sum_{\alpha=1}^{m} \eta_{\alpha\alpha} \partial_k X^\alpha * \partial_i \partial_j X^\alpha. \tag{4.8}$$

It is straightforward that the metric g given by (4.5) satisfies (4.1) and the chiral coefficients Υ_{ijk} given by (4.6) satisfy (4.2). Hence in this case, the two Ricci curvatures satisfy (4.3) and (4.4).

Inspired by (3.1), the following noncommutative Einstein field equations

$$R_j^i + \Theta_j^i - \delta_j^i R = T_j^i$$

were proposed in [5]. As these equations may not capture all information of noncommutative metrics, the following strong version

$$R_j^i - \frac{1}{2}\delta_j^i R = T_j^i, \quad \Theta_j^i - \frac{1}{2}\delta_j^i R = \tilde{T}_j^i$$

was given in [23]. Theorem 4.1 indicates that only the first one is sufficient and the noncommutative Einstein field equations should be

$$R_j^i - \frac{1}{2}\delta_j^i R = T_j^i$$

if (4.1), (4.2) hold, in particular, if noncommutative metrics are given by isometric embedding.

5　Quantum effects of gravity

Noncommutative geometric quantities $g_{ij}, R_{lkij}, R_j^i, R, \cdots$ are by definition formal power series whose coefficients are smooth functions. These power series are usually divergent. But if they are given by quantum fluctuation of certain spherically symmetric isometric embeddings, then they are absolutely convergent and have closed forms. This indicates that the quantization of gravity is renormalizable. A typical spherically symmetric spacetime is the Schwarzschild metric

$$g_{Sch} = -\left(1 - \frac{2m}{r}\right)dt^2 + \frac{dr^2}{1 - \dfrac{2m}{r}} + r^2\left(d\theta^2 + \sin^2\theta d\psi^2\right).$$

There are two ways to embed it into higher dimensional flat spaces. One is the Kasner's embedding [10] into $\mathbb{R}^{2,4}$

$$X^1 = \left(1 - \frac{2m}{r}\right)^{\frac{1}{2}}\sin t,$$

$$X^2 = \left(1 - \frac{2m}{r}\right)^{\frac{1}{2}}\cos t,$$

$$X^3 = f(r), \quad (f')^2 + 1 = \left(1 - \frac{2m}{r}\right)^{-1}\left(1 + \frac{m^2}{r^4}\right),$$

$$X^4 = r\sin\theta\cos\phi,$$

$$X^5 = r\sin\theta\sin\phi,$$

$$X^6 = r\cos\theta.$$

Another is the Fronsdal's embedding [7] into $\mathbb{R}^{1,5}$

$$X^1 = \left(1 - \frac{2m}{r}\right)^{\frac{1}{2}}\sinh t,$$

$$X^2 = \left(1 - \frac{2m}{r}\right)^{\frac{1}{2}}\cosh t,$$

$$X^3 = f(r), \quad (f')^2 + 1 = \left(1 - \frac{2m}{r}\right)^{-1}\left(1 - \frac{m^2}{r^4}\right),$$

$$X^4 = r\sin\theta\cos\phi,$$

$$X^5 = r\sin\theta\sin\phi,$$

$$X^6 = r\cos\theta.$$

With coordinates $(x^1, x^2, x^3, x^4) = (t, r, \theta, \phi)$, choose the matrix (θ^{ij}) with only nonzero items

$$\theta^{34} = -\theta^{43} = 1$$

to define the Moyal product. The above two isometric embeddings give rise to the same quantum fluctuation [19]

$$g_{11} = -\left(1 - \frac{2m}{r}\right),$$

$$g_{12} = g_{21} = g_{13} = g_{31} = g_{14} = g_{41} = 0,$$

$$g_{22} = \left(1 - \frac{2m}{r}\right)^{-1}\left[1 + \left(1 - \frac{2m}{r}\right)(\sin^2\theta - \cos^2\theta)\sinh^2\hbar\right],$$

$$g_{23} = g_{32} = 2r\sin\theta\cos\theta\sinh^2\hbar,$$

$$g_{24} = -g_{42} = -2r\sin\theta\cos\theta\sinh\hbar\cosh\hbar,$$

$$g_{33} = r^2\left[1 - (\sin^2\theta - \cos^2\theta)\sinh^2\hbar\right],$$

$$g_{34} = -g_{43} = r^2\left(\sin^2\theta - \cos^2\theta\right)\sinh\hbar\cosh\hbar,$$

$$g_{44} = r^2\left[\sin^2\theta + (\sin^2\theta - \cos^2\theta)\sinh^2\hbar\right].$$

The noncommutative scalar curvature is [19]

$$R = \frac{A(r,\theta,\hbar)}{B(r,\theta,\hbar)},$$

$$A(r,\theta,0) = 0, \quad A(0,\theta,\hbar) \neq 0, \quad A(2m,\theta,\hbar) \neq 0,$$

$$B(r,\theta,\hbar) = 8r^2\Big[3r + 10m + (r - 2m)(4\cosh 2\hbar + \cosh 4\hbar)$$

$$- 16m\cos 2\theta\sinh^2\hbar\Big]^3.$$

Then we observe that the noncommutative Schwarzschild metric still has a black hole with the event horizon at $r = 2m$. Since both g and R do not depend on the time t, this quantum black hole cannot be evaporated.

In general, by calculating Moyal products of trigonometric functions, we obtain the following theorem in [8].

Theorem 5.1 ([8])　*Let the open set*

$$U = (0, \infty) \times (0, 2\pi) \times (0, \pi) \times \cdots \times (0, \pi) \subset \mathbb{R}^n,$$

which is equipped with coordinates $(x^1, x^2, \cdots, x^n) = (\rho, \theta_1, \cdots, \theta_{n-1})$. *Let* $(U, g[0])$ *be a (pseudo-) Riemannian metric given by a spherically symmetric isometric embedding*

$$X : U \longrightarrow \mathbb{R}^{p,m-p}$$

with

$$X^1 = f^1(\rho),$$

$$\cdots\cdots$$

$$X^{m-n} = f^{m-n}(\rho),$$

$$X^{m-n+1} = f^{m-n+1}(\rho) \sin\theta_{n-1} \sin\theta_{n-2} \cdots \sin\theta_2 \sin\theta_1,$$

$$X^{m-n+2} = f^{m-n+2}(\rho) \sin\theta_{n-1} \sin\theta_{n-2} \cdots \sin\theta_2 \cos\theta_1,$$

$$\cdots\cdots$$

$$X^{m-2} = f^{m-2}(\rho) \sin\theta_{n-1} \sin\theta_{n-2} \cos\theta_{n-3},$$

$$X^{m-1} = f^{m-1}(\rho) \sin\theta_{n-1} \cos\theta_{n-2},$$

$$X^m = f^m(\rho) \cos\theta_{n-1},$$

where $f^1(\rho), \cdots, f^m(\rho)$ are smooth functions of ρ, $m - n + 1 > p$ and

$$f^{m-n+1}(\rho) = f^{m-n+2}(\rho).$$

Fix some $l \in [3, n]$, define the Moyal product in terms of skew-symmetric matrix (θ^{ij}) with nonzero elements

$$\theta^{2l} = -\theta^{l2} = \lambda \neq 0.$$

Then the quantum fluctuation of $g[0]$ and their curvatures have closed forms coming from absolutely convergent power series expansions on U.

Another surprising result in the study of quantum effects of gravity is the discovery of exact solutions of noncommutative vacuum Einstein field equations, which are quantum fluctuations of plane-fronted gravitational waves [20].

The noncommutative vacuum field equations are

$$R_j^i = \Theta_j^i = 0.$$

Recall that the plane-fronted gravitational waves have the following metric

$$g_{pp} = dx^2 + dy^2 + 2dudv + 2H(x, y, u)du^2,$$
$$R_j^i = 0 \iff H_{xx} + H_{yy} = 0.$$

The metric has additivity

$$(H_i)_{xx} + (H_i)_{yy} = 0 \implies \left(\sum H_i\right)_{xx} + \left(\sum H_i\right)_{yy} = 0.$$

The metric g_{pp} can be isometrically embedded in the 6-dimensional flat space with signature $(4, -2)$ [6],

$$X = \left(x, y, \frac{Hu + u + v}{\sqrt{2}}, \frac{Hu - \frac{u^2}{2}}{\sqrt{2}}, \frac{Hu - u + v}{\sqrt{2}}, \frac{Hu + \frac{u^2}{2}}{\sqrt{2}}\right).$$

Define the Moyal product by an arbitrary constant skew-symmetric matrix (θ^{ij}). The quantum fluctuation g given by X is [20]

$$g_{xx} = g_{yy} = g_{uv} = g_{vu} = 1,$$

$$g_{xy} = g_{yx} = g_{xv} = g_{vx} = g_{yv} = g_{vy} = 0,$$

$$g_{xu} = -g_{ux} = -\hbar(\theta_{yu}H_{xy} + \theta_{xu}H_{xx}),$$

$$g_{yu} = -g_{uy} = -\hbar(\theta_{yu}H_{yy} + \theta_{xu}H_{xy}),$$

$$g_{uu} = 2H$$

and the Ricci curvatures are

$$R_3^4 = \Theta_3^4 = -H_{xx} - H_{yy}, \quad \text{others} = 0.$$

Hence it is an exact vacuum solution of noncommutative Einstein field equations if and only if

$$H_{xx} + H_{yy} = 0.$$

The additivity also holds for the quantum metric g.

6   Quasi-connections and curvatures

Let $B_q(\cdot, \cdot)$ be a family of bi-differential operators on M. A star product on M is an associative $\mathbb{R}[[\hbar]]$-bilinear product

$$\star : \mathcal{A}_M \times \mathcal{A}_M \longrightarrow \mathcal{A}_M$$

such that, for $u,\ v \in C^\infty(M)$,

$$1 \star u = u \star 1 = u, \quad u \star v = uv + \sum_{q=1}^{\infty} B_q(u,v)\hbar^q.$$

The associativity of $\star$ implies that

$$\{u,v\}_\star = \frac{1}{2}\Big(B_1(u,v) - B_1(v,u)\Big)$$

is a Poisson bracket, cf. [11, 12]. If the skew-symmetric matrix (θ^{ij}) is not constant, Moyal product is not associative. However, on a Poisson manifold, Kontsevich proved that there always exists a star product $\star$ such that $\{\cdot,\cdot\}_\star$ gives rise to the Poisson structure [11]. Such a star product is called a deformation quantization of the Poisson manifold [4, 11]. Since the coefficients of a star product are bi-differential operators, it can be restricted to $\mathcal{A}_U$.

We now fix a star product $\star$. The noncommutative left (resp. right) $\star$-tangent bundle ${}^\star\mathcal{T}_U$ (resp. ${}^\star\tilde{\mathcal{T}}_U$) on U is defined to be the free left (resp. right) $(\mathcal{A}_U, \star)$-module with basis $\{E_1, \cdots, E_n\}$ (resp. $\{\tilde{E}_1, \cdots, \tilde{E}_n\}$), i.e.,

$${}^\star\mathcal{T}_U = \Big\{ a^i \star E_i \,\Big|\, a^i \in \mathcal{A}_U,\ a^i \star E_i = 0 \Longleftrightarrow a^i = 0 \Big\},$$

$${}^\star\tilde{\mathcal{T}}_U = \Big\{ \tilde{E}_i \star \tilde{a}^i \,\Big|\, \tilde{a}^i \in \mathcal{A}_U,\ \tilde{E}_i \star \tilde{a}^i = 0 \Longleftrightarrow \tilde{a}^i = 0 \Big\}.$$

An element of ${}^\star\mathcal{T}_U$ (resp. ${}^\star\tilde{\mathcal{T}}_U$) is called a left (resp. right) $\star$-vector field. A noncommutative $\star$-metric ${}^\star g$ is defined as a homomorphism of two-sided $(\mathcal{A}_U, \star)$-modules

$$ {}^\star g : {}^\star\mathcal{T}_U \otimes_{\mathbb{R}[[\hbar]]} {}^\star\tilde{\mathcal{T}}_U \longrightarrow \mathcal{A}_U$$

such that the matrix

$$({}^\star g_{ij}) \in \mathcal{A}_U^{n\times n}, \quad {}^\star g_{ij} = {}^\star g(E_i, \tilde{E}_j)$$

is $\star$-invertible, i.e., there exists a unique matrix $({}^\star g^{ij}) \in \mathcal{A}_U^{n\times n}$ such that

$$ {}^\star g_{ik} \star {}^\star g^{kj} = {}^\star g^{jk} \star {}^\star g_{ki} = \delta_i^j.$$

Same as that for the Moyal product in [8], we can show that, on U,

$$({}^\star g_{ij}) \text{ is invertible} \Longleftrightarrow ({}^\star g_{ij}[0]) \text{ is invertible}.$$

In terms of noncommutative $\star$-metric $^\star g$, dual bases E^i, $\tilde{E}^j$ of $\tilde{E}_j$, E_i respectively can be induced via

$$^\star g(E^i, \tilde{E}_j) = {}^\star g(E_j, \tilde{E}^i) = \delta^i_j.$$

The noncommutative left (resp. right) $\star$-cotangent bundle $^\star \mathcal{T}^*_U$ (resp. $^\star \tilde{\mathcal{T}}^*_U$) on U with respect to the noncommutative metric $^\star g$ is the free left (resp. right) $(\mathcal{A}_U, \star)$-module with basis $\{E^1, \cdots, E^n\}$ (resp. $\{\tilde{E}^1, \cdots, \tilde{E}^n\}$)

$$^\star \mathcal{T}^*_U = \left\{ a_i \star E^i \,\middle|\, a_i \in \mathcal{A}_U,\, a_i \star E^i = 0 \Longleftrightarrow a_i = 0 \right\},$$

$$^\star \tilde{\mathcal{T}}^*_U = \left\{ \tilde{E}^i \star \tilde{a}_i \,\middle|\, \tilde{a}_i \in \mathcal{A}_U,\, \tilde{E}^i \star \tilde{a}_i = 0 \Longleftrightarrow \tilde{a}_i = 0 \right\}.$$

As star products are not compatible with the Leibniz rule, connections can not be defined in general. But we can still define quasi-connections as well as their curvatures. A noncommutative left (resp. right) quasi-connection $^\star \nabla$ is a map

$$^\star \nabla : {}^\star \mathcal{T}_U \longrightarrow {}^\star \tilde{\mathcal{T}}^*_U \otimes_{\mathcal{A}_U} {}^\star \mathcal{T}_U \quad (\text{resp. } {}^\star \tilde{\nabla} : {}^\star \tilde{\mathcal{T}}_U \longrightarrow {}^\star \tilde{\mathcal{T}}_U \otimes_{\mathcal{A}_U} {}^\star \mathcal{T}^*_U)$$

such that noncommutative left (resp. right) covariant derivatives

$$^\star \nabla_i : {}^\star \mathcal{T}_U \longrightarrow {}^\star \mathcal{T}_U \quad (\text{resp. } {}^\star \tilde{\nabla}_i : {}^\star \tilde{\mathcal{T}}_U \longrightarrow {}^\star \tilde{\mathcal{T}}_U),$$

defined by

$$^\star \nabla_i V = {}^\star g(E_i, \tilde{E}^k) \star W_k \quad (\text{resp. } {}^\star \tilde{\nabla}_i \tilde{V} = \tilde{W}_k \star {}^\star g(E^k, \tilde{E}_i))$$

for any

$$^\star \nabla V = \tilde{E}^k \otimes W_k,\ W_k \in {}^\star \mathcal{T}_U \quad (\text{resp. } {}^\star \tilde{\nabla} \tilde{V} = \tilde{W}_k \otimes E^k,\ \tilde{W}_k \in {}^\star \tilde{\mathcal{T}}^*_U),$$

are $\mathbb{R}[[\hbar]]$-linear. Quasi-connections are not compatible with the Leibniz rule and can not be determined completely by connection coefficients.

Curvature operators $^\star \mathcal{R}_{E_i E_j}$ and $^\star \tilde{\mathcal{R}}_{\tilde{E}_i \tilde{E}_j}$ for quasi-connections can be defined in the same way as in Section 2. But they are not $(\mathcal{A}_U, \star)$-linear. The corresponding left (resp. right) Riemannian curvatures, Ricci curvatures and scalar curvatures are formally defined as follows [8].

$$^\star R_{lkij} = {}^\star g(^\star \mathcal{R}_{E_i E_j} E_k, \tilde{E}_l),$$

$$^\star \tilde{R}_{lkij} = -\,{}^\star g(E_k, {}^\star \tilde{\mathcal{R}}_{\tilde{E}_i \tilde{E}_j} \tilde{E}_l),$$

$$^\star R_{kj} = {}^\star R_{lkij} \star {}^\star g^{li}, \quad {}^\star \Theta_{il} = {}^\star g^{jk} \star {}^\star R_{lkij},$$

$$^\star \tilde{R}_{kj} = {}^\star \tilde{R}_{lkij} \star {}^\star g^{li}, \quad {}^\star \tilde{\Theta}_{il} = {}^\star g^{jk} \star {}^\star \tilde{R}_{lkij},$$

$$^\star R_j^p = {}^\star g^{pk} \star {}^\star R_{lkij} \star {}^\star g^{li}, \quad {}^\star \Theta_i^p = {}^\star g^{jk} \star {}^\star R_{lkij} \star {}^\star g^{lp},$$

$$^\star \tilde{R}_j^p = {}^\star g^{pk} \star {}^\star \tilde{R}_{lkij} \star {}^\star g^{li}, \quad {}^\star \tilde{\Theta}_i^p = {}^\star g^{jk} \star {}^\star \tilde{R}_{lkij} \star {}^\star g^{lp},$$

$$^\star R = {}^\star R_j^j = {}^\star \Theta_i^i, \quad {}^\star \tilde{R} = {}^\star \tilde{R}_j^j = {}^\star \tilde{\Theta}_i^i.$$

As the metric compatibility does not hold for quasi-connections, the left and right Riemannian curvatures are not equal to each other. It was shown in [8] that the first (algebraic) Bianchi identity

$$^\star \mathcal{R}_{E_i E_j} E_k + {}^\star \mathcal{R}_{E_j E_k} E_i + {}^\star \mathcal{R}_{E_k E_i} E_j = 0,$$

$$^\star \tilde{\mathcal{R}}_{E_i E_j} E_k + {}^\star \tilde{\mathcal{R}}_{E_j E_k} E_i + {}^\star \tilde{\mathcal{R}}_{E_k E_i} E_j = 0$$

holds if the quasi-connection is torsion-free, i.e.,

$$^\star \nabla_i E_j = {}^\star \nabla_j E_i, \quad (\text{resp. } {}^\star \tilde{\nabla}_i \tilde{E}_j = {}^\star \tilde{\nabla}_j \tilde{E}_i).$$

An isometric embedding

$$X : (U, g[0]) \longrightarrow \mathbb{R}^{p, m-p}$$

also yields a noncommutative $\star$-metric and a quasi-connection. For $Y = (Y^1, \cdots, Y^m)$, $Z = (Z^1, \cdots, Z^m) \in \mathcal{A}_U^m$, we denote

$$Y \star_\eta Z = \sum_{\alpha=1}^m \eta_{\alpha\alpha} Y^\alpha \star Z^\alpha.$$

Then the $\star$-metric is

$$^\star g(E_i, \tilde{E}_j) = \partial_i X \star_\eta \partial_j X.$$

For the construction of the quasi-connection, it was shown in [8] that the left (resp. right) $(\mathcal{A}_U, \star)$-module homomorphism

$$\sigma : {}^\star\mathcal{T}_U \longrightarrow \mathcal{A}_U^m \quad (\text{resp. } \tilde{\sigma} : {}^\star\tilde{\mathcal{T}}_U \longrightarrow \mathcal{A}_U^m)$$

given by

$$\sigma(E_i) = \partial_i X, \quad (\text{resp. } \tilde{\sigma}(\tilde{E}_i) = \partial_i X)$$

is injective and induces a direct sum decomposition

$$\mathcal{A}_U^m = \sigma({}^\star\mathcal{T}_U) \oplus {}^\star\mathcal{N}_U \quad (\text{resp. } \mathcal{A}_U^m = \tilde{\sigma}({}^\star\tilde{\mathcal{T}}_U) \oplus {}^\star\tilde{\mathcal{N}}_U) \tag{6.1}$$

of left (resp. right) modules, where

$$^\star\mathcal{N}_U = \left\{ Y \in \mathcal{A}_U^m \middle| Y \star_\eta \tilde{\sigma}(\tilde{E}_i) = 0, \forall i \right\},$$

$$^\star\tilde{\mathcal{N}}_U = \left\{ Y \in \mathcal{A}_U^m \middle| \sigma(E_i) \star_\eta Y = 0, \forall i \right\}.$$

Denote

$$\mathrm{pr}_1 : \mathcal{A}_U^m \longrightarrow \sigma({}^\star\mathcal{T}_U) \quad (\text{resp. } \tilde{\mathrm{pr}}_1 : \mathcal{A}_U^m \longrightarrow \tilde{\sigma}({}^\star\tilde{\mathcal{T}}_U))$$

the projection onto the first factor with respect to the decomposition (6.1). The left (resp. right) quasi-connection

$$^\star\nabla : {}^\star\mathcal{T}_U \longrightarrow {}^\star\tilde{\mathcal{T}}_U^* \otimes_{\mathcal{A}_U} {}^\star\mathcal{T}_U \quad (\text{resp. } {}^\star\tilde{\nabla} : {}^\star\tilde{\mathcal{T}}_U \longrightarrow {}^\star\tilde{\mathcal{T}}_U \otimes_{\mathcal{A}_U} {}^\star\mathcal{T}_U^*)$$

is given by

$$^\star\nabla V = \tilde{E}^k \otimes {}^\star\nabla_k V \quad (\text{resp. } {}^\star\tilde{\nabla}\tilde{V} = {}^\star\tilde{\nabla}_k\tilde{V} \otimes E^k) \tag{6.2}$$

for any $V \in {}^\star\mathcal{T}_U$ (resp. $\tilde{V} \in {}^\star\tilde{\mathcal{T}}_U$), where the left (resp. right) quasi-covariant derivative

$$^\star\nabla_i : {}^\star\mathcal{T}_U \longrightarrow {}^\star\mathcal{T}_U \quad (\text{resp. } {}^\star\tilde{\nabla}_i : {}^\star\tilde{\mathcal{T}}_U \longrightarrow {}^\star\tilde{\mathcal{T}}_U),$$

for each i, is given by

$$^\star\nabla_i(V) = \sigma^{-1}\big(\mathrm{pr}_1(\partial_i\sigma(V))\big) \quad \big(\text{resp. } {}^\star\tilde{\nabla}_i(\tilde{V}) = \tilde{\sigma}^{-1}\big(\tilde{\mathrm{pr}}_1(\partial_i\tilde{\sigma}(\tilde{V}))\big)\big). \tag{6.3}$$

By (6.3) and

$$\mathrm{pr}_1\big(\partial_i\sigma(E_j)\big) = \mathrm{pr}_1(\partial_i\partial_j X), \quad \tilde{\mathrm{pr}}_1\big(\partial_i\tilde{\sigma}(\tilde{E}_j)\big) = \tilde{\mathrm{pr}}_1(\partial_i\partial_j X),$$

we know that the quasi-connection (6.2) is torsion-free.

References

[1] P. Aschieri, C. Blohmann, M. Dimitrijevic, et al., *A gravity theory on noncommutative spaces*, Class. Quant. Grav., 22(17): 3511, 2005.

[2] P. Aschieri, M. Dimitrijevic, F. Meyer, et al., *Noncommutative geometry and gravity*, Class. Quant. Grav., 23(6): 1883, 2006.

[3] F. Bayen, M. Flato, C. Fronsdal, et al., *Quantum mechanics as a deformation of classical mechanics*, Lett. Math. Phys., 1: 521–530, 1977.

[4] F. Bayen, M. Flato, C. Fronsdal, et al., *Deformation theory and quantization. I. Deformations of symplectic structures*, Ann. Phys., 111(1): 61–110, 1978.

[5] M. Chaichian, A. Tureanu, R. B. Zhang, et al., *Riemannian geometry of noncommutative surfaces*, J. Math. Phys., 49(7): 073511, 2008.

[6] C. D. Collinson, *Embeddings of the plane-fronted waves and other space-times*, J. Math. Phys., 9(3): 403–410, 1968.

[7] C. Fronsdal, *Completion and embedding of the Schwarzschild solution*, Phys. Rev., 116(3): 778, 1959.

[8] H. Gao, X. Zhang, *Deformation quantization and intrinsic noncommutative differential geometry*, arXiv:2304.03581, 2023.

[9] M. Gerstenhaber, *On deformation of rings and algebras*, Ann. Math., 79(1): 59–103, 1964.

[10] E. Kasner, *Finite representation of the solar gravitational field in flat space of six dimensions*, Amer. J. Math., 43(2): 130–133, 1921.

[11] M. Kontsevich, *Deformation quantization of Poisson manifolds*, Lett. Math. Phys., 66: 157–216, 2003.

[12] C. Laurent-Gengoux, A. Pichereau, P. Vanhaecke, *Poisson Structures*, Springer Science & Business Media, 2012.

[13] J. Lukierski, M. Woronowicz, *Spinorial Snyder and Yang models from superalgebras and noncommutative quantum superspaces*, Phys. Lett. B, 824: 136783, 2022.

[14] J. Lukierski, M. Woronowicz, *Noncommutative spaces and superspaces from Snyder and Yang type models*, arXiv:2204.07787, 2022.

[15] M. Pavšič, V. Tapia, *Resource letter on geometrical results for embeddings and branes*, arXiv: gr-qc/0010045, 2000.

[16] N. Seiberg, E. Witten, *String theory and noncommutative geometry*, J. High Energy Phys., 1999(9): 032, 1999.

[17] H. S. Snyder, *Quantized space-time*, Phys. Rev., 71(1): 38–41, 1947.

[18] A. Touati, S. Zaim, *Geodesic equation in non-commutative gauge theory of gravity*, Chinese Phys. C, 46(10): 105101, 2022.

[19] D. Wang, R. B. Zhang, X. Zhang, *Quantum deformations of Schwarzschild and Schwarzschild-de Sitter spacetimes*, Class. Quant. Grav. 26(8): 085014, 2009.

[20] D. Wang, R. B. Zhang, X. Zhang, *Exact solutions of noncommutative vacuum Einstein field equations and plane-fronted gravitational waves*, Eur. Phys. J. C, 64: 439–444, 2009.

[21] C. N. Yang, *On quantized space-time*, Phys. Rev., 72(9): 874, 1947.

[22] R. B. Zhang, X. Zhang, *Projective module description of embedded noncommutative spaces*, Rev. Math. Phys., 22(5): 507–531, 2010.

[23] X. Zhang, *Deformation quantization and noncommutative black holes*, Sci. China Math., 54: 2501–2508, 2011.

The Isoperimetric Problem in Riemannian Manifolds Endowed with Conformal Vector Fields

JIAYU LI
SHUJING PAN
School of Mathematical Sciences, University of Science and Technology of China
E-mail: jiayuli@ustc.edu.cn; psj@ustc.edu.cn

Abstract This is a survey of the results in [14] regarding the isoperimetric problem in the Riemannian manifold. We consider a mean curvature type flow in the Riemannian manifold endowed with a non-trivial conformal vector field, which was firstly introduced by Guan and Li [8] in space forms. This flow preserves the volume of the bounded domain enclosed by a star-shaped hypersurface and decreases the area of hypersurface under certain conditions. We will prove the long time existence and convergence of the flow. As a result, the isoperimetric inequality for such a domain is established.

1 Background

The isoperimetric problem is one of the most famous problems in geometry with a long history. Many methods were introduced to study the isoperimetric problem. For example, ABP technique, optimal transport, Knothe mapping and so on. Recently, Brendle [2] obtained a sharp isoperimetric inequality for minimal submanifolds in Euclidean space of codimension at most 2 by proving the Sobolev inequality on submanifolds.

It has been proven that the hypersurface flow is a valid tool to study the isoperimetric problem in manifolds. Gage and Hamilton [6] proved the isoperimetric inequality for convex planar domains using the curve shortening flow. Huisken [11] introduced the volume preserving mean curvature flow and proved the isoperimetric inequality for closed, uniformly convex hypersurfaces in $\mathbb{R}^{n+1}$. Applying the volume preserving mean curvature flow in [11], Huisken and Yau [12] showed that if M is a C^4-asymptotic flat 3-manifold, then the complement of a compact subset of M admitting a foliation by a family of stable constant mean curvature spheres. Moreover, the leaves of this foliation are the unique stable CMC spheres within a large class of surfaces. After that, their uniqueness result was strengthened in the important work by Qing and Tian [17]. Eichmair and Metzger [5] confirmed that the surfaces found in [12,17] are in fact isoperimetric surfaces (conjectured by Bray [1] first).

Besides the volume preserving mean curvature flow, there are other mean curvature type flows which can be used to study the isoperimetric problem. Schulze [19] proved the isoperimetric inequality in $\mathbb{R}^{n+1}$ with $n \leqslant 7$ using a nonlinear mean curvature type flow. Guan and Li [8] constructed a new mean curvature type flow to prove the isoperimetric inequality for star-shaped hypersurfaces in space forms. After that, Guan, Li and Wang [10] extended the mean curvature type flow to a class of warped product spaces and proved the isoperimetric inequality for star-shaped hypersurfaces under certain conditions.

In this paper, we are interested in the isoperimetric problem in the Riemannian manifold endowed with a non-trivial conformal vector field. In section 2, we review the flow constructed by Guan and Li [8] in space forms (Guan, Li and Wang [10] in warped product spaces). In section 3, we introduce some preliminaries about the conformal vector field. In section 4, we apply the Guan-Li's flow to solve the isoperimetric problem for star-shaped hypersurfaces in the Riemannian manifold endowed with a non-trivial conformal vector field.

2　Guan-Li's mean curvature type flow

Let $(\mathbf{B}^n, \tilde{g})$ be a closed Riemannian manifold, $\phi = \phi(r)$ be a smooth positive function defined on the interval $[r_0, \bar{r}]$ for $r_0 < \bar{r}$. Consider a Riemannian manifold $(\mathbf{N}^{n+1}, \bar{g})$ with the warped product structure

$$\bar{g} = dr^2 + \phi^2 \tilde{g}, \quad r \in [r_0, \bar{r}], \tag{2.1}$$

where $\tilde{g}$ is the metric of the manifold $\mathbf{B}^n$. Especially, for space forms, $(B^n, \tilde{g})$ is the standard unit sphere $\mathbb{S}^n \subset \mathbb{R}^{n+1}$ and

$$\phi(r) = \begin{cases} r, \ r \in [0, +\infty), \ \text{when } \mathbf{N}^{n+1} = \mathbb{R}^{n+1}, \\ \sin r, \ r \in [0, \pi), \ \text{when } \mathbf{N}^{n+1} = \mathbb{S}^{n+1}, \\ \sinh r, \ r \in [0, +\infty), \ \text{when } \mathbf{N}^{n+1} = \mathbb{H}^{n+1}. \end{cases}$$

Then, there is a natural closed conformal vector field $V = \phi \partial_r$, which means

Lemma 1 (*cf. [8, Lemma 2.1]*)   *In $(\mathbf{N}^{n+1}, \bar{g})$, the vector field $V = \phi \partial_r$ satisfies*

$$\overline{\nabla}_i V_j = \phi'(r) \bar{g}_{ij}, \tag{2.2}$$

where $\overline{\nabla}$ is the covariant derivative with respect to the metric $\bar{g}$.

Let Σ be a smooth closed embedded hypersurface in $(\mathbf{N}^{n+1}, \bar{g})$, σ_k be the k-th elementary symmetric functions of the principal curvature. Using the above lemma, the following important Minkowski identities hold:

Proposition 2 (*cf. [10]*)   *Let Σ be a closed hypersurface embedded in $\mathbf{N}^{n+1}$, we have*

$$\int_\Sigma n\phi' - uH d\mu = 0,$$

and

$$(n-1) \int_\Sigma \phi' H d\mu = 2 \int_\Sigma \sigma_2 u d\mu + \int_\Sigma u \left(\overline{Ric}(\partial_r, \partial_r) - \overline{Ric}(\nu, \nu) \right) d\mu,$$

where ν is the unit outward normal, H is the mean curvature and $u = \bar{g}(\phi \partial_r, \nu)$ is the support function of Σ.

Depending on these features, Guan and Li [8] constructed a mean curvature type flow in space forms, and they also generalized it to warped product spaces together with Wang [10]. They considered the following evolution equation for a family of embeddings $F(\cdot, t)$ of hypersurfaces:

$$\frac{\partial F}{\partial t} = (n\phi' - uH)\nu. \tag{2.3}$$

Applying the Minkowski identities given in Proposition 2, they showed that the flow (2.3) has an important property for hypersurfaces which are star-shaped, i.e $u = \bar{g}(\phi \partial_r, \nu) > 0$.

Theorem 3 (*cf. [10, Theorem 2.7]*) *Let Σ_t be the smooth, star-shaped solution of the flow (2.3) in the warped product space $(\mathbf{N}^{n+1}, \bar{g})$. Denote by Ω_t the domain enclosed by the hypersurface Σ_t. If $\phi(r)$ and $\tilde{g}$ satisfy the conditions:*

$$\tilde{Ric} \geqslant (n-1)K\tilde{g}, \quad (\phi')^2 - \phi\phi'' \leqslant K, \tag{2.4}$$

where $K > 0$ is a constant and $\tilde{Ric}$ is the Ricci curvature of $\tilde{g}$. Then, along the flow (2.3), the volume of Ω_t is a constant and the area of Σ_t is non-increasing.

Next, they proved the long time existence and convergence of the flow (2.3). Since the star-shaped hypersurface can be expressed as a radial graph over $\mathbf{B}^n$ in $\mathbf{N}^{n+1}$ as

$$\Sigma = \{(r(\theta), \theta) \in [r_0, \bar{r}) \times \mathbf{B}^n\},$$

where $r(\theta)$ is a smooth and positive function on $\mathbf{B}^n$. The flow (2.3) is equivalent to the following quasilinear parabolic initial value problem:

$$\partial_t r(\theta, t) = (n\phi' - uH)\frac{\omega}{\phi}, \tag{2.5}$$

where $\omega = \sqrt{1 + |\tilde{\nabla} r|^2}$. Therefore, the convergence theorem follows from the parabolic estimates of r.

Theorem 4 (*cf. [10, Theorem 1.1]*) *Let Σ_0 be a smooth graphical hypersurface in $\mathbf{N}^{n+1}$. If $\phi(r)$ and $\tilde{g}$ satisfy the conditions:*

$$\tilde{Ric} \geqslant (n-1)K\tilde{g}, \quad 0 \leqslant (\phi')^2 - \phi\phi'' \leqslant K, \tag{2.6}$$

where $K > 0$ is a constant and $\tilde{Ric}$ is the Ricci curvature of $\tilde{g}$. Then, the evolution equation (2.3) with Σ_0 as initial data has a smooth graphical solution for $t \in [0, +\infty)$. Moreover, the solution hypersurfaces converge exponentially to a level set of r as $t \to \infty$.

As an application, they obtained a solution to the isoperimetric problem in warped product spaces.

Theorem 5 (*cf. [10, Theorem 1.2]*) *Assume $\phi(r)$ and $\tilde{g}$ satisfy the conditions in Theorem 4. Then, given the enclosed volume, the slice $\{r\} \times \mathbf{B}^n$ minimizes the area among all star-shaped hypersurfaces.*

The flow (2.3) was also generalized to the fully-nonlinear version by Guan and Li [9] in the Euclidean space and by Chen, Guan, Li and Scheuer

[3] in the sphere, by replacing the mean curvature H to general curvature functions $\sigma_{k+1}/\sigma_k, k = 1, \ldots, n$. There are many other interesting results concerning the locally constrained curvature flows to study Alexandrov-Fenchel type inequalities in space forms, such as [4, 13, 18] among many others.

3   Conformal vector fields

We shall use the Guan-Li's mean curvature type flow (2.3) to solve the isoperimetric problem in more general ambient manifolds besides the warped product space. More precisely, we consider the Riemannian manifold admitting a non-trivial conformal vector field.

Definition 6   Let $(M^{n+1}, \bar{g})$ be a Riemannian manifold, a vector field ξ on M is called a conformal vector field if it satisfies

$$L_\xi \bar{g}_{ij} = \overline{\nabla}_i \xi_j + \overline{\nabla}_j \xi_i = 2\varphi \bar{g}_{ij},$$

where L is the Lie derivative and φ is given by $\varphi = \dfrac{div\xi}{n+1}$. Moreover, ξ is called closed if it satisfies

$$\overline{\nabla}_i \xi_j = \varphi \bar{g}_{ij}.$$

Note that if M is the warped product space, then $\xi = V = \phi \partial_r$ is a special closed conformal vector field (recall (2.2)).

Let $\mathcal{Z}(\xi)$ be the set which consists of zero points of the vector field ξ, it is also called the singular set. If $\varphi > 0$ in M, we can conclude that $\mathcal{Z}(\xi)$ is a discrete set by Obata's results in [16]. Let $M' = M \setminus \mathcal{Z}(\xi)$ be the open dense set and $\mathcal{F}(\xi)$ be the foliation in M' determined by the n-dimensional distribution

$$p \in M' \to \mathcal{D}(p) = \{v \in T_p M | \langle \xi(p), v \rangle = 0\}.$$

When ξ is closed, the distribution $\mathcal{D}$ is integrable on M'. Montiel proved that the following property:

Proposition 7 (cf. [15, Prop 1])   Let $(M^{n+1}, \bar{g})$ be a Riemannian manifold endowed with a closed, conformal vector field ξ. Then, we have that each connected leaf of $\mathcal{F}(\xi)$ is a totally umbilical hypersurface with constant mean curvature. Moreover, the functions $|\xi|$, $div\xi$ and $\xi(\varphi)$ are constant on connected leaves of $\mathcal{F}(\xi)$.

In addition, Montiel [15] claimed that a Riemannian manifold admitting a non-trivial closed, conformal vector field is locally isometric to a warped product with a 1-dimensional factor. Therefore, in such a manifold, based on the isoperimetric inequality in warped product space which was established by Guan, Li and Wang [10], one might be wondering whether similar result still holds. We confirmed this fact as follows:

Theorem 8   *Let $(M^{n+1}, \bar{g})$ be a Riemannian manifold admitting a non-trivial closed, conformal vector field ξ. We assume further that M satisfies the following conditions:*

(i) $\varphi > 0$;

(ii) $\varphi^2 - \xi(\varphi) \geqslant 0$;

(iii) the direction determined by ξ is of least Ricci curvature on M', i.e

$$\overline{Ric}(\mathcal{N}, \mathcal{N}) \leqslant \overline{Ric}(e, e), \quad \forall e \in TM, \ |e| = 1,$$

where $\mathcal{N} = \dfrac{\xi}{|\xi|}$.

Then, given the enclosed volume, the leaf of the foliation $\mathcal{F}(\xi)$ minimizes the area among all star-shaped hypersurfaces.

It's easy to check that our condition (iii) is a weaker assumption than the condition (2.4) used in [10].

Next, we wonder whether the leaf of the foliation $\mathcal{F}$ is the solution of isoperimetric problem in general Riemannian manifolds admitting non-trivial conformal vector fields without the closed assumption of ξ. We observed the following property:

Proposition 9   *Let $(M^{n+1}, \bar{g})$ be a Riemannian manifold endowed with a conformal vector field ξ. Assume that the n-dimensional distribution $\mathcal{D}$ defined on M' is integrable and determines a Riemannian foliation $\mathcal{F}(\xi)$ whose connected leaf is a closed hypersurface oriented by the unit vector field $\mathcal{N} = \dfrac{\xi}{|\xi|}$, then each connected leaf of $\mathcal{F}(\xi)$ is totally umbilical and has mean curvature $H = n\dfrac{\varphi}{|\xi|}$.*

From the above proposition, we notice that any leaf of $\mathcal{F}(\xi)$ is totally umbilical but may not has constant mean curvature. On the other hand, the first variation formula of the area implies that an isoperimetric hypersurface must have constant mean curvature. Hence, in the rest of this paper, we assume that any leaf of $\mathcal{F}(\xi)$ is a closed hypersurface with constant mean curvature. Under this condition, we proved that the classical Minkowski identities given in Proposition 2 still hold in M, which plays a crucial role in the later proof.

Proposition 10   *Let $(M^{n+1}, \bar{g})$ be a Riemannian manifold endowed with a complete conformal vector field ξ. We assume further that any connected leaf of the foliation $\mathcal{F}(\xi)$ has constant mean curvature. Then on the closed hypersurface $\Sigma \subset M'$ we have,*

$$\int_{\Sigma} n\varphi - uH d\mu = 0, \tag{3.1}$$

and

$$(n-1) \int_{\Sigma} \varphi \sigma_1 d\mu = 2 \int_{\Sigma} \sigma_2 u d\mu + \int_{\Sigma} u \left(\overline{Ric}(\mathcal{N}, \mathcal{N}) - \overline{Ric}(\nu, \nu) \right) d\mu. \tag{3.2}$$

where ν is the unit normal vector field of Σ and $u = \langle \xi, \nu \rangle$ is the support function.

4   Main results and the strategy of proofs

In this section, we are prepared to answer the question:

Whether the leaf of the foliation $\mathcal{F}$ is the solution of isoperimetric problem in the general Riemannian manifold $(M^{n+1}, \bar{g})$ admitting a non-trivial conformal vector field?

We assume that $(M^{n+1}, \bar{g})$ satisfies the following conditions:

(i) $\varphi > 0$;

(ii) $\varphi^2 - \xi(\varphi) > 0$;

(iii) any connected leaf of totally umbilical Riemannian foliation $\mathcal{F}(\xi)$ on M' is a closed hypersurface with constant mean curvature and it can be represented by the level set $\left\{ \frac{|\xi|}{\varphi} = r \right\}$ for some constant $r > 0$;

(iv) the direction determined by ξ is of least Ricci curvature on M', that is

$$\overline{Ric}(\mathcal{N}, \mathcal{N}) \leqslant \overline{Ric}(e, e), \quad \forall e \in TM, \ |e| = 1.$$

Applying the Minkowski identities we obtained in Proposition 10, we generalize the Guan-Li's mean curvature type flow in $(M^{n+1}, \bar{g})$. We consider the following evolution equation for a family of embeddings $F(\cdot, t)$ of hypersurfaces in $(M^{n+1}, \bar{g})$:

$$\frac{\partial F}{\partial t} = (n\varphi - uH)\nu, \tag{4.1}$$

where $u = \bar{g}(\xi, \nu)$ is the support function.

Firstly, we prove the monotone property of the flow (4.1):

Theorem 11   *Assume that $(M^{n+1}, \bar{g})$ is a Riemannian manifold admitting a non-trivial conformal vector field ξ which satisfies the conditions (i)-(iv). Let Σ_t be the smooth, star-shaped solution of the flow (4.1) in $(M^{n+1}, \bar{g})$. Denote by Ω_t the domain enclosed by the hypersurface Σ_t. Then, along the flow (4.1), the volume of Ω_t is a constant and the area of Σ_t is non-increasing.*

Proof   The monotone property of the flow (4.1) is a consequence of the Minkowski identities. By direct calculation, using (3.1) and (3.2), along the flow (4.1) we have

$$\frac{d}{dt} \text{vol}(\Omega_t) = \int_{\Sigma_t} n\varphi - uH = 0,$$

and

$$\frac{d}{dt} \text{area}(\Sigma_t)$$
$$= \int_{\Sigma} \left(\frac{2n}{n-1}\sigma_2 - H^2 + \frac{n}{n-1}(\overline{Ric}(\mathcal{N}, \mathcal{N}) - \overline{Ric}(\nu, \nu)) \right) u d\mu \leqslant 0 \tag{4.2}$$

$\square$

Next, with a star-shaped initial data Σ_0, we want to prove that the solution hypersurfaces Σ_t of the flow (4.1) are still star-shaped and exist for all time $t \in [0, +\infty)$ and converge to a smooth closed hypersurface Σ_∞. Then, combining the monotone property, it's convinced that Σ_∞ is a solution of the isoperimetric problem. Unlike the warped product space case, Σ_t can't be expressed as a radial graph now, and the flow (4.1) is failed to be converted into a scalar parabolic equation as (2.5). Fortunately, we find a function can play a similar role in our case as the radial function in the warped product space. Precisely, on a hypersurface $\Sigma \subset M'$, choosing the local normal coordinate $\{e_i\}_{i=1}^n$, we have

$$\nabla_i \frac{|\xi|^2}{\varphi^2} = 2\frac{\varphi^2 - \xi(\varphi)}{\varphi^3} \langle \xi, e_i \rangle,$$

and

$$
\nabla_j \nabla_i \frac{|\xi|^2}{\varphi^2} = \frac{2}{\varphi^2 |\xi|^2} \xi \left(\frac{\varphi^2 - \xi(\varphi)}{\varphi} \right) \langle \xi, e_i \rangle \langle \xi, e_j \rangle - 2 \frac{\varphi^2 - \xi(\varphi)}{\varphi^4} e_i(\varphi) \langle \xi, e_j \rangle
$$
$$
- 2 \frac{\varphi^2 - \xi(\varphi)}{\varphi^4} e_j(\varphi) \langle \xi, e_i \rangle + 2 \frac{\varphi^2 - \xi(\varphi)}{\varphi^2} g_{ij} - 2 \frac{\varphi^2 - \xi(\varphi)}{\varphi^3} u h_{ij}.
$$

Hence, along the flow (4.1), the a priori estimates of the function $\dfrac{|\xi|^2}{\varphi^2}$ on Σ_t implies the a priori estimates of the second fundamental forms of Σ_t in M which leads to the long-time existence and convergence.

Applying the parabolic maximum principle, we have the following C^0 and C^1 estimates of $\dfrac{|\xi|^2}{\varphi^2}$.

Theorem 12   *Let Σ_0 be a closed hypersurface embedded in M', if $\Sigma(t)$ is a smooth solution of the flow (4.1) starting from Σ_0. Then, $\forall (p,t) \in \Sigma \times [0,T)$ we have*

$$
\min_{p \in \Sigma} \frac{|\xi|^2}{\varphi^2}(p,0) \leqslant \frac{|\xi|^2}{\varphi^2}(p,t) \leqslant \max_{p \in \Sigma} \frac{|\xi|^2}{\varphi^2}(p,0).
$$

From this, we conclude that $\Sigma(t) \cap \mathcal{Z}(\xi) = \emptyset$ along the flow.

The above theorem guarantees Σ_t always lies in a compact domain surrounded by two leaves $\left\{ \dfrac{|\xi|^2}{\varphi^2} = \min\limits_{p \in \Sigma} \dfrac{|\xi|^2}{\varphi^2}(p,0) \right\}$ and $\left\{ \dfrac{|\xi|^2}{\varphi^2} = \max\limits_{p \in \Sigma} \dfrac{|\xi|^2}{\varphi^2}(p,0) \right\}$ of the foliation $\mathcal{F}(\xi)$.

The C^1 estimate of $\dfrac{|\xi|^2}{\varphi^2}$ is equivalent to the uniform lower bound of the support function u, which also implies that the flow (4.1) preserves the star-shaped property.

Theorem 13   *Let Σ_0 be a star-shaped, closed hypersurface embedded in M'. Then along the flow (4.1), there exists a constant $\epsilon > 0$ independent of the time t such that*

$$
\min_{t \geqslant 0} u \geqslant \epsilon.
$$

By direct computation, along the flow (4.1), we can see that the function $\dfrac{|\xi|^2}{\varphi^2}$ evolves by

$$\partial_t \frac{|\xi|^2}{\varphi^2} = u\Delta_g \frac{|\xi|^2}{\varphi^2} - 2\frac{u}{\varphi^2|\xi|^2}\xi\left(\frac{\varphi^2 - \xi(\varphi)}{\varphi}\right)(|\xi|^2 - u^2)$$
$$+ 4\frac{\varphi^2 - \xi(\varphi)}{\varphi^4}e_i(\varphi)\langle\xi, e_i\rangle u,$$

this is a qusilinear parabolic equation. The above C^0 and C^1 estimates of $\frac{|\xi|^2}{\varphi^2}$ imply that this equation is uniformly parabolic. Then the regularity of function $\frac{|\xi|^2}{\varphi^2}$ follows from the standard parabolic theory. Now, we can conclude that the norm of the second fundamental form and its higher derivatives of the solution hypersurfaces Σ_t have uniform bounds that are independent of the time t along the flow (4.1). Then, by Arzela-Ascoli theorem, for any time sequence $t_j \to \infty$ we know that there exists a subsequence (still denoted by t_j) such that Σ_{t_j} converges smoothly to a limit hypersurface Σ_∞. We have the following conclusion:

Theorem 14 ([14]) *Let Σ_0 be a star-shaped, closed hypersurface embedded in M'. Then the evolution equation (2.3) with Σ_0 as an initial data has a smooth solution for $t \in [0, +\infty)$. Moreover, any sequence of the solution hypersurfaces has a convergent subsequence, and each limit hypersurface is a totally umbilical hypersurface whose unit normal vector field ν_∞ attains the least Ricci curvature on M', that means*

$$\overline{Ric}(\nu_\infty, \nu_\infty) = \overline{Ric}(\mathcal{N}, \mathcal{N}).$$

Proof The characterization of the limit hypersurface follows from the monotonicity formula (4.2). □

Finally, we confirm that the limit hypersurface is just the leaf of the foliation $\mathcal{F}(\xi)$ which encloses a domain whose volume is equal to the enclosed domain of Σ_0. The first result is as follows:

Theorem 15 *Assume that $(M^{n+1}, \bar{g})$ is a Riemannian manifold admitting a non-trivial conformal vector field ξ which satisfies the conditions (i)-(iv). In addition, we assume that our hypothesis on the Ricci curvature of M is strict, i.e*

$$\overline{Ric}(\mathcal{N}, \mathcal{N}) < \overline{Ric}(e, e), \quad \forall e \in TM, |e| = 1, e \neq \mathcal{N}. \tag{4.3}$$

Let Σ_0 be a closed hypersurface embedded in M', if $\Sigma(t)$ is a smooth solution of the flow (4.1) starting from Σ_0. Then, the limit hypersurface Σ_∞ is a leaf of the foliation $\mathcal{F}(\xi)$.

If we don't assume that the hypothesis on the Ricci curvature of M is strict, then we need another assumption on the sectional curvature of M.

Theorem 16 *Assume that $(M^{n+1}, \bar{g})$ is a Riemannian manifold admitting a non-trivial conformal vector field ξ which satisfies the conditions (i)-(iv). In addition, we assume that the sectional curvature of the direction ξ satisfies*

$$K(X, \xi) \geqslant -\frac{\varphi^2}{|\xi|^2}, \quad \forall X \in TM, \ X \neq \xi, \tag{4.4}$$

where $K(X, \xi) = \dfrac{\bar{R}(X, \xi, X, \xi)}{|X|^2 |\xi|^2 - \langle X, \xi \rangle^2}$. Let Σ_0 be a closed hypersurface embedded in M', if $\Sigma(t)$ is a smooth solution of the flow (4.1) starting from Σ_0. Then, the limit hypersurface Σ_∞ is a leaf of the foliation $\mathcal{F}(\xi)$.

Consequently, we have proved that the leave of the foliation $\mathcal{F}(\xi)$ is the solution of the isoperimetric problem in M with extra condition (4.3) or (4.4) among all star-shaped hypersurfaces.

Theorem 17 *Assume that $(M^{n+1}, \bar{g})$ is a Riemannian manifold admitting a non-trivial conformal vector field ξ which satisfies the conditions (i)-(iv). In addition, we assume that our hypothesis on the Ricci curvature of M is strict, or the sectional curvature of the direction ξ satisfies (4.4). Then, given the enclosed volume, the leaf of the foliation $\mathcal{F}(\xi)$ minimizes the area among all star-shaped hypersurfaces.*

In a recent work [7], Flynn and Reznikov claimed that our extra assumption (4.3) or (4.4) is unnecessary, the limit hypersurface Σ_∞ is a leaf of the foliation $\mathcal{F}(\xi)$ under the conditions (i)-(iv). They also released the star-shaped condition to a weaker version.

References

[1] H. Bray, *The Penrose inequality in general relativity and volume comparison theorems involving scalar curvature*, PhD thesis, Stanford University, 1997.

[2] S. Brendle, *The isoperimetric inequality for a minimal submanifold in Euclidean space*, J. Amer. Math. Soc., 34(2): 595–603, 2021.

[3] C. Chen, P. Guan, J.F. Li, J. Scheuer, *A fully-nonlinear flow and quermassintegral inequalities in the sphere*, Pure Appl. Math. Q, 18(2): 437–461, 2022.

[4] M. Chen, J. Sun, *Alexandrov-Fenchel type inequalities in the sphere*, Adv. Math, 397, Paper No. 108203, 25pp, 2022.

[5] M. Eichmair, J. Metzger, *Large isoperimetric surfaces in initial data sets*, J. Differ. Geom., 94(1):159–186, 2013.

[6] M. Gage, R. Hamilton, *The heat equation shrinking convex plane curves*, J. Differ. Geom., 23(1): 69–96, 1986.

[7] J. Flynn, J. Reznikov, *General conformally induced mean curvature flow*, arXiv: 2309.14679, 2023.

[8] P. Guan, J. F. Li, *A mean curvature type flow in space forms*, Int. Math. Res. Not., 13: 4716–4740, 2015.

[9] P. Guan, J.F. Li, *A fully-nonlinear flow and quermassintegral inequalities (in Chinese)*, Sci. Sin. Math, 48: 147–156, 2018.

[10] P. Guan, J. F. Li, M. Wang, *A volume preserving flow and the isoperimetric problem in warped product spaces*, Trans. Amer. Math. Soc., 372(4): 2777–2798, 2019.

[11] G. Huisken, *The volume preserving mean curvature flow*, J. Reine Angew. Math., 382: 35–48, 1987.

[12] G. Huisken, S. Yau, *Definition of center of mass for isolated physical systems and unique foliations by stable spheres with constant mean curvature*, Inven. Math., 124(1-3): 281–311, 1996.

[13] Y. Hu, H. Li, Y. Wei, *Locally constrained curvature flows and geometric inequalities in hyperbolic space*, Math. Ann., 382(3-4): 1425-1474, 2022.

[14] J. Y. Li, S. Pan, *The isoperimetric problem in the Riemannian manifold admitting a non-trivial conformal vector field*, arXiv:2303.17887, 2023.

[15] S. Montiel, *Unicity of constant mean curvature hypersurfaces in some Riemannian manifolds*, Indiana Univ. Math. J., 48: 711–748, 1999.

[16] M. Obata, *Conformal transformations of Riemannian manifolds*, J. Differ. Geom., 4(3): 311–333, 1970.

[17] J. Qing, G. Tian, *On the uniqueness of the foliation of spheres of constant mean curvature in asymptotically flat 3-manifolds*, J. Amer. Math. Soc., 20(4):1091–1110, 2007.

[18] J. Scheuer, C. Xia, *Locally constrained inverse curvature flows*, Trans. Amer. Math. Soc, 372(10): 6771–6803, 2019.

[19] F. Schulze, *Nonlinear evolution by mean curvature and isoperimetric inequalities*, J. Differ. Geom., 79(2): 197–241, 2008.

Adiabatic Limit Formula for Bismut-Cheeger Eta Forms

Bo Liu

School of Mathematical Sciences, Key Laboratory of Mathematics and Engineering Applications(Ministry of Education) & Shanghai Key Laboratory of PMMP, East China Normal University, Shanghai, China

E-mail: bliu@math.ecnu.edu.cn

Mengqing Zhan

School of Science & Big Data Science, Zhejiang University of Science and Technology, Hangzhou, China

E-mail: mqzhan@zust.edu.cn

Abstract   This is a survey of the results in [10, 13–15] considering the generalization of adiabatic limit formula in [11] to the family case.

1   Introduction

The famous Atiyah-Singer index theorem unifies the Chern-Gauss-Bonnet theorem, the Hirzebruch-Riemann-Roch theorem and the Hirzebruch signature theorem. Its original version concerns a compact manifold without boundary, which is also called a closed manifold (cf. [2]). After that, there are three generalizations we are interested in.

The first one is for manifold with boundary. Then the Atiyah-Patodi-Singer (APS) eta invariant constructed from the Dirac operator on the boundary appears as the boundary correction term for the index formula under a global boundary condition (cf. [1]). The second is for manifold with a compact Lie group action. The resulting equivariant index formula leads to various interesting results of Lie groups and representation theory. The third one concerns the fibration of compact manifolds, which leads to the family index theorem.

It's natural to combine them. The extension of the APS index theorem is a story about the extension of the eta invariant. The family version of APS index theorem is established in [7, 8, 17, 18], where the family extension of eta invariant η is the Bismut-Cheeger eta form $\widetilde{\eta}$, which serves as the boundary correction term in the family index theorem for manifolds with boundary. The equivariant APS index theorem is studied by H. Donnelly [12], who extends the eta invariant η to the equivariant version η_g and shows that it contributes the boundary term of the equivariant index theorem. The equivariant version of the family case leads naturally to the definition of the equivariant eta form $\widetilde{\eta}_g$ (see (2.10)).

Let's say something more about the discovery of the Bismut-Cheeger eta form. For a fibration of closed manifolds $\pi_X : Z \to Y$ with closed fiber X, we can define a Dirac operator D_T on Z under some condition, where T is a metric-rescaled parameter for the fiber X. The adiabatic limit formula for eta invariant is a way of calculating the limit $\lim_{T \to +\infty} \eta(D_T)$. It turns out that we can investigate the asymptotic behavior of the spectrum of D_T, decomposing it by different orders of $1/T$. Originated in [20], E. Witten studies this adiabatic limit when the base manifold is a circle for physical consideration and calls it global anomaly. Later in [6, 11], J.-M. Bismut and J. Cheeger illustrate the 0-order part is a combination of geometry and a differential form, named after Bismut-Cheeger eta form. This is how the eta form arises in history. The 1-order part is eta invariant related to another Dirac operator for S. The higher order parts are recognized in [11], in which these parts are interpreted as the signature of Leray spectral sequences in the special case of signature operators.

Since the eta invariant has the family extension, it is reasonable to expect that all index theoretic results involving eta invariants have family version with eta invariants replaced by Bismut-Cheeger eta forms. In fact, we can further take the group action into consideration and study the equivariant results. This survey serves to explain how the adiabatic limit formula for eta invariants of (2.4) extends to the equivariant eta forms following this thread.

This survey is organized as follows. Section 2 is devoted to introducing the eta invariant for Dirac operator and adiabatic limit formula of eta invariant. Then we represent how the adiabatic limit formula of eta invariant generalizes to the equivariant Bismut-Cheeger eta form in Section 3.

In this survey, all manifolds are supposed to be closed and oriented. We assume that all fibrations here are submersions with closed oriented fibers. We denote by d the exterior differential operator and d^S when we like to insist the base manifold S. We use the Einstein summation convention in this paper: when an index variable appears twice in a single term and is not otherwise defined, it implies summation of that term over all the values of the index.

2   Adiabatic limit

In this section, we will recall the classical adiabatic limit formula for eta invariants and the definition of the equivariant Bismut-Cheeger eta form.

2.1   Adiabatic limit for eta invariant

In this subsection, we will figure out the geometry to define an eta invariant and write down the adiabatic limit formula for eta invariants in [6] and [11].

Let X be a closed oriented manifold with Riemannian metric g^{TX}. Let $\mathrm{Cl}(TX)$ be the Clifford algebra bundle of (TX, g^{TX}), whose fibre at $x \in X$ is the Clifford algebra $\mathrm{Cl}(T_x X)$ of the Euclidean vector space $(T_x X, g^{T_x X})$. A Clifford module $\mathcal{E}$ is a complex vector bundle over X equipped with a Hermitian metric $h^{\mathcal{E}}$ and a Clifford multiplication c of $\mathrm{Cl}(TX)$ on $\mathcal{E}$ such that the action c restricted to TX is skew-adjoint on $(\mathcal{E}, h^{\mathcal{E}})$. Let $\nabla^{\mathcal{E}}$ be a Clifford connection on $\mathcal{E}$, which is a connection preserving h^E and parallel with the Clifford action. Then the Dirac operator D on $\mathcal{E}$ is defined by

$$D = \sum_{i=1}^{\dim X} c(e_i) \nabla_{e_i}^{\mathcal{E}}, \qquad (2.1)$$

where $\{e_i\}$ is a locally oriented orthonormal basis of TX [3, §3.3]. The eta invariant is defined by

$$\eta(D) := \int_0^\infty \mathrm{Tr}[D \exp(-tD^2)] \frac{dt}{\sqrt{\pi t}}. \qquad (2.2)$$

Now we present the adiabatic limit formula. Let $\pi_X : Z \to Y$ be a fibration with fiber X. Let $TX := \ker(d\pi_X : TZ \to TY)$ be the relative bundle over Z.

Let g^{TX} be the Euclidean metric on TX. Let $\mathcal{E}_X$ be a complex vector bundle over Z with Hermitian metric $h^{\mathcal{E}_X}$ such that when restricted to $Z_y, y \in Y$, $\mathcal{E}_X|_{Z_y}$ is a Clifford module of $\mathrm{Cl}(TX)$. We say $\mathcal{E}_X$ is a Clifford module of $\mathrm{Cl}(TX)$. Let $\nabla^{\mathcal{E}_X}$ be a Clifford connection on $\mathcal{E}_X$. We denote by $\underline{\mathcal{E}_X} := (\mathcal{E}_X, h^{\mathcal{E}_X}, \nabla^{\mathcal{E}_X})$ a geometric triple and we can define the fiberwise Dirac operator $D^{\mathcal{E}_X}$ as in (2.1). Usually we also use the notation D^X for simplicity.

Let $T_X^H Z$ be a horizontal subbundle of TZ. Then we have $TZ = T_X^H Z \oplus TX$ and $T_X^H Z \simeq \pi_X^* TY$. We denote by $\pi_X := (\pi_X, T_X^H Z, g^{TX})$. Let g^{TY} be a Riemannian metric on TY. We set

$$g_T^{TZ} := \pi_X^* g^{TY} \oplus \frac{1}{T^2} g^{TX}. \tag{2.3}$$

Let $\mathcal{E}_Y$ be a Clifford module over Y with Hermitian metric $h^{\mathcal{E}_Y}$. Let $\nabla^{\mathcal{E}_Y}$ be a Clifford connection with respect to $(\mathcal{E}_Y, h^{\mathcal{E}_Y})$. Then $\mathcal{E} := \pi_X^* \mathcal{E}_Y \otimes \mathcal{E}_X$ is a Clifford module over Z and we can construct a Clifford connection $\nabla^{\mathcal{E},T}$ associated with g_T^{TZ} (cf. [13, (4.3)]) and the corresponding Dirac operator D_T^Z.

We assume that $\ker D^X$ forms a vector bundle over Y and denote by $D^{\mathcal{E}_Y \otimes \ker D^X}$ the corresponding Dirac operator. The adiabatic limit formula is as follows.

Theorem 1 ([6,11]) *We assume that $\dim \ker D_T^Z$ is constant for T large enough, then*

$$\lim_{T \to +\infty} \eta(D_T^Z) = 2 \int_Y \widehat{A}(TY) \mathrm{ch}(\mathcal{E}_Y/S) \widetilde{\eta}(\pi_X, \underline{\mathcal{E}_X}) + \eta(D^{\mathcal{E}_Y \otimes \ker D^X}) + R. \tag{2.4}$$

Here $\widetilde{\eta}(\pi_X, \underline{\mathcal{E}_X}) \in \Omega^\bullet(Y)$ is the famous Bismut-Cheeger eta form we will discuss later (cf. (2.10)), $\widehat{A}(TY)$ and $\mathrm{ch}(\mathcal{E}_Y/S)$ are $\widehat{A}$-form and relative Chern character form respectively (cf. [3, §1.5]) and $R \in \mathbb{Z}$ is a remainder term. Remark that $\widetilde{\eta}(\pi_X, \underline{\mathcal{E}_X})$ satisfies the following transgression formula:

$$d^Y \widetilde{\eta}(\pi_X, \underline{\mathcal{E}_X}) = \begin{cases} \int_X \widehat{A}(TX) \mathrm{ch}(\mathcal{E}_Y/S) - \mathrm{ch}(\ker D^{\mathcal{E}_X}), & \dim X \text{ is even;} \\ \int_X \widehat{A}(TX) \mathrm{ch}(\mathcal{E}_Y/S), & \dim X \text{ is odd.} \end{cases} \tag{2.5}$$

If Y is a point and $\dim X$ is odd, $\widetilde{\eta}(\pi_X, \underline{\mathcal{E}_X}) = \frac{1}{2} \eta(D^{\mathcal{E}_X})$.

Let's explain the remainder term R. In [6], if Z and Y are spin, for invertible D^Y, $R = 0$. In [11], for the general case, $R = \sum_{\lambda_0, \lambda_1 = 0} \mathrm{sgn}(\lambda_T) \in \mathbb{Z}$. Let's talk about it in more detail. Suppose that λ_T is an eigenvalue of D_T^Z, which is bounded when $T \to +\infty$. As $T \to +\infty$, λ_T has the asymptotic expansion of $1/T$ with $\lambda_T \sim \lambda_k T^{-k} + \lambda_{k+1} T^{-(k+1)} + \cdots$, $\lambda_k \neq 0$. Let $\mathrm{sgn}(\lambda_T)$ be the sign of λ_k. Thus $\sum_{\lambda_0, \lambda_1 = 0} \mathrm{sgn}(\lambda_T)$ is the sum of the signs of those eigenvalues which have $\lambda_0 = \lambda_1 = 0$ in the asymptotic expansion.

We expect that if there is a family of fibrations $\pi_X : Z \to Y$ over another manifold S, we can get a family of Dirac operators D_T^Z over S, which can be used to establish the family version of (2.4) with all eta invariants replaced by eta forms. Such formula will be useful in differential K-theory.

2.2 Definition of the equivariant Bismut-Cheeger eta form

In this subsection, we will introduce the definition of the equivariant Bismut-Cheeger eta form. We use the same notation as the last subsection.

Let $P^{TX} : TZ \to TX$ be the orthogonal projection with respect to g_1^{TZ}. Take ∇^{TY}, ∇^{TZ} the corresponding Levi-Civita connections. We define $\nabla^{TX} := P^{TX} \nabla^{TZ} P^{TX}$. Let $\{e_i\}_{i=1}^{\dim X}, \{f_p\}_{p=1}^{\dim Y}$ be local orthonormal basis of TX, TY respectively with duals $\{e^i\}_{i=1}^{\dim X}, \{f^p\}_{p=1}^{\dim Y}$. For $A \in TY$, we write A^H for the horizontal lift of A in $T_X^H Z$. Let $c(\mathcal{T}) = -\frac{1}{2} c(P^{TX}[f_p^H, f_q^H]) f^p \wedge f^q \wedge$. Let $\mathscr{E}_b := C^\infty(X_b, \mathcal{E}), b \in Y$, then $\{\mathscr{E}_b\}_{b \in Y}$ form an infinite dimensional vector bundle over Y, which is equipped with the natural inner product $\langle \cdot, \cdot \rangle_{\mathscr{E}_b} := \int_{X_b} h^{\mathcal{E}}(\cdot, \cdot) dv_X$. We could define a connection $\nabla^{\mathscr{E}}$ on $\mathscr{E}$ as in [9, Proposition 1.4] which preserves $\langle \cdot, \cdot \rangle_{\mathscr{E}}$. The following rescaled Bismut superconnection is proposed by Bismut in [5]:

$$B_t = \sqrt{t} D^X + \nabla^{\mathscr{E}} - \frac{1}{4\sqrt{t}} c(\mathcal{T}), \quad t > 0. \tag{2.6}$$

Let G be a compact Lie group acting smoothly on Z. We assume that G commutes with π_X, acts on Y trivially and preserves all geometric data. In particular, we assume that the group action commutes with the Dirac operators.

Let P be a trace class operator on $\Lambda(T^*Y) \widehat{\otimes} \mathrm{End}(\mathscr{E})$ which takes values in $\Lambda(T^*Y)$. We denote by $\mathrm{Tr}^{\mathrm{odd/even}}[P]$ the part of $\mathrm{Tr}_s[P]$ which takes values in odd or even forms. Set

$$
\widetilde{\mathrm{Tr}}[P] := \begin{cases} \mathrm{Tr}_s[P], & \text{if } \dim X \text{ is even;} \\ \mathrm{Tr}^{\mathrm{odd}}[P], & \text{if } \dim X \text{ is odd.} \end{cases} \tag{2.7}
$$

Here $\mathrm{Tr}_s[P]$ denotes the supertrace of P as in [3, §1.3]. For $\alpha \in \Omega^i(Y)$, set

$$
\psi_Y(\alpha) = \begin{cases} \left(\dfrac{1}{2\pi\sqrt{-1}}\right)^{\frac{i}{2}} \cdot \alpha, & i \text{ is even;} \\[2mm] \dfrac{1}{\sqrt{\pi}} \left(\dfrac{1}{2\pi\sqrt{-1}}\right)^{\frac{i-1}{2}} \cdot \alpha, & i \text{ is odd,} \end{cases} \tag{2.8}
$$

and

$$
\widetilde{\psi}_Y(\alpha) = \begin{cases} \dfrac{1}{\sqrt{\pi}} \left(\dfrac{1}{2\pi\sqrt{-1}}\right)^{\frac{i}{2}} \cdot \alpha, & i \text{ is even;} \\[2mm] \left(\dfrac{1}{2\pi\sqrt{-1}}\right)^{\frac{i+1}{2}} \cdot \alpha, & i \text{ is odd.} \end{cases} \tag{2.9}
$$

For $\beta \in \Omega^\bullet(Y \times [0,1]_u)$, if we write $\beta = \beta_0 + du \wedge \beta_1$, with $\beta_0, \beta_1 \in \Omega(T^*Y)$, we set $[\beta]^{du} := \beta_1$.

For $g \in G$, let Z^g be the fixed point set of g-action on Z. Then $\pi|_{X^g} : Z^g \to Y$ is a fiber bundle with fiber X^g. We assume that TX^g is oriented.

Definition 2 ([13, Definition 2.3]) For $g \in G$, the *equivariant Bismut-Cheeger eta form* $\widetilde{\eta}_g(\pi_X, \mathcal{E}_X) \in \Omega^\bullet(Y)$ is defined by

$$
\widetilde{\eta}_g(\pi_X, \mathcal{E}_X)
$$

$$
:= -\int_0^{+\infty} \left\{ \psi_Y \widetilde{\mathrm{Tr}}\left[g \exp\left(-\left(B_u + du \wedge \frac{\partial}{\partial u} \right)^2 \right) \right] \right\}^{du} du
$$

$$
= \begin{cases} \displaystyle\int_0^{+\infty} \widetilde{\psi}_Y \mathrm{Tr}^{\mathrm{even}}\left[g \frac{\partial B_u}{\partial u} \exp(-B_u^2) \right] du \in \Omega^{\mathrm{even}}(B;\mathbb{C}), & \text{if } \dim X \text{ is odd;} \\[4mm] \displaystyle\int_0^{+\infty} \widetilde{\psi}_Y \mathrm{Tr}_s\left[g \frac{\partial B_u}{\partial u} \exp(-B_u^2) \right] du \in \Omega^{\mathrm{odd}}(B;\mathbb{C}), & \text{if } \dim X \text{ is even.} \end{cases} \tag{2.10}
$$

Remark that the original Bismut-Cheeger eta form in (2.4), $\widetilde{\eta}(\pi_X, \mathcal{E}_X) = \widetilde{\eta}_e(\pi_X, \mathcal{E}_X)$, where e is the unity of G.

3 Adiabatic limit formula for eta forms

In this section, we will discuss the generalizations of the adiabatic limit formula.

3.1 Generalizations

In this subsection, we will analyze the geometry of the composition of fibrations to state the family extension of adiabatic limit formula covering the works of [10, 13–15].

Let $\pi_X : W \to V$, $\pi_Y : V \to S$ be two successive equivariant fibrations with fibers X and Y respectively. We call this the composition of fibrations. Let $\pi_Z := \pi_Y \circ \pi_X : W \to S$ be the composed equivariant fibration with fiber Z. Note that π_X induces another fibration $\pi_X : Z \to Y$ with fiber X. We assume that the group action on S is trivial. We can depict the composition as follows:

$$
\begin{array}{ccc}
X \longrightarrow Z \longrightarrow W & & \\
\Big\downarrow{\scriptstyle \pi_X} \quad \Big\downarrow{\scriptstyle \pi_X} \quad {\scriptstyle \pi_Z} & & \\
Y \longrightarrow V \xrightarrow{\ \pi_Y\ } S.
\end{array}
\tag{3.1}
$$

Let $\underline{\pi_X} = (\pi_X, T_X^H W, g^{TX})$, $\underline{\pi_Y} = (\pi_Y, T_Y^H V, g^{TY})$ and $\underline{\pi_Z} = (\pi_Z, T_Z^H W, g^{TZ})$ be equivariant geometric data with respect to π_X, π_Y and π_Z. Assume that $T_Z^H W \subset T_X^H W$ and $g^{TZ} = \pi_X^* g^{TY} \oplus g^{TX}$. Let ∇^{TX}, ∇^{TY} and ∇^{TZ} be the corresponding connections on TX, TY and TZ. Set ${}^0\nabla^{TZ} := \pi_X^* \nabla^{TY} \oplus \nabla^{TX}$.

Let $\underline{\mathcal{E}_X} = (\mathcal{E}_X, h^{\mathcal{E}_X}, \nabla^{\mathcal{E}_X})$ (resp. $\underline{\mathcal{E}_Y} = (\mathcal{E}_Y, h^{\mathcal{E}_Y}, \nabla^{\mathcal{E}_Y})$) be a G-equivariant self-adjoint Clifford module of $\mathrm{Cl}(TX)$ over W (resp. Clifford module of $\mathrm{Cl}(TY)$ over V) with a G-invariant Clifford connection. Let $\mathcal{E}_Z = \pi_X^* \mathcal{E}_Y \otimes \mathcal{E}_X$. Then $\underline{\mathcal{E}_Z} = (\mathcal{E}_Z, \pi_X^* h^{\mathcal{E}_Y} \otimes h^{\mathcal{E}_X}, \nabla^{\mathcal{E}})$ is a $\mathbb{Z}_2$-graded G-equivariant self-adjoint $\mathrm{Cl}(TZ)$-module over W with a G-invariant Clifford connection. Here $\nabla^{\mathcal{E}}$ is defined in [13, (4.3)]. We assume that all fixed point sets are oriented.

Theorem 3 ([10, 13–15]) *For $g \in G$, modulo exact forms on S,*

$$
\tilde{\eta}_g(\underline{\pi_Z}, \underline{\mathcal{E}}) = \int_{Y^g} \widehat{A}_g(TY) \mathrm{ch}_g(\mathcal{E}_Y/S) \tilde{\eta}_g(\pi_X^g, \underline{\mathcal{E}_X})
$$

$$
- \int_{Z^g} \widetilde{\widehat{A}}_g(TZ, \nabla^{TZ}, {}^0\nabla^{TZ}) \mathrm{ch}_g(\mathcal{E}_Z/S) + \tilde{\eta}_g(\underline{\pi_Y}, \mathcal{E}_Y \otimes \ker D^{\mathcal{E}_X}) + \widetilde{R}.
$$

$$
\tag{3.2}
$$

Here $\widehat{A}_g(TZ, {}^0\nabla^{TZ})$ and $\mathrm{ch}_g(\mathcal{E}/S)$ are the equivariant characteristic forms (see [3, §7.1] for the definitions), $\widetilde{\widehat{A}}_g(TZ, \nabla^{TZ}, {}^0\nabla^{TZ})$ is the Chern-Simons form of the equivariant $\widehat{A}$-forms $\widehat{A}_g(TZ, \nabla^{TZ})$ and $\widehat{A}_g(TZ, {}^0\nabla^{TZ})$.

Compared with the equation (2.4), all eta invariants have been extended to Bismut-Cheeger eta forms. The limit process in (2.4) is reformulated by the second term on the right hand side of (3.2) [15, Remark 1.2]. The last term $\widetilde{R}$ should be the family extension of R in (2.4).

There are some works which are related to the equation (3.2). In [10], the authors show that in the case of signature operators on flat bundles, without group action, the term $\widetilde{R}$ is the sum of finite dimensional eta forms constructed by spectral sequence, which can degenerate to the remainder term in (2.4). In [13], the author establishes the equation (3.2) in the case of $\ker D^Z = 0$. In this case, $\widetilde{R} = 0$. A more recent work is [14], in which the author proves the general case of Theorem 3 with $\widetilde{R}$ belongs to $\mathrm{ch}_g(K_G^0(S))$. However, in this situation, $\widetilde{R}$ can't be recognized as the family extension of R. In [15], under some mild assumptions (see [15, Assumption 3.7]), we characterize $\widetilde{R}$ explicitly.

In the following subsections, we will explain some details in [15] for the construction of $\widetilde{R}$ by finite dimensional eta forms.

3.2   Finite dimensional eta form

The finite dimensional eta form comes from the finite version of the family index theorem. The objective of this subsection is to define the equivariant finite dimensional eta form.

Let (E, h^E) be a G-equivariant $\mathbb{Z}_2$-graded Hermitian vector bundle over M with Hermitian connection ∇^E. Let V be a G-equivariant endomorphism of E super-commuting with $\Omega^\bullet(M)$. Then $E' = \ker V$ is a G-equivariant $\mathbb{Z}_2$-graded bundle with the projected geometric data $h^{E'} := P^{\ker V} h^E P^{\ker V}$, $\nabla^{E'} := P^{\ker V} h^E P^{\ker V}$.

Quillen in [19] defines the superconnection, which is a prototype of Bismut superconnection, as follows:

$$L := \nabla^E + V. \tag{3.3}$$

Compared with (2.6), we define the rescaled superconnection

$$L_t := \nabla^E + \sqrt{t}V. \tag{3.4}$$

Quillen studies the supertrace of the heat operator of L_t, $\mathrm{Tr}_s[\exp(-L_t^2)]$, to prove that $\mathrm{ch}(E, \nabla^E)$ and $\mathrm{ch}(E', \nabla^{E'})$ are in the same cohomology class. In fact, this can be extended to the equivariant case naturally.

For $g \in G$, to measure the difference between $\mathrm{ch}_g(E', \nabla^{E'})$ and $\mathrm{ch}_g(E, \nabla^E)$ as differential forms, we could define the equivariant finite dimensional eta form as

$$\widetilde{\eta}_g(E, E', \nabla^E, \nabla^{E'}) := -\int_0^{+\infty} \widetilde{\psi}_M \mathrm{Tr}_s \left[g \frac{\partial L_t}{\partial t} \exp(-L_t^2) \right] dt. \tag{3.5}$$

We call it the finite dimensional eta form as it shares the same transgression property of Bismut-Cheeger eta form (2.5):

Proposition 4 *Modulo exact form on M, we have*

$$d\widetilde{\eta}_g(E, E', \nabla^E, \nabla^{E'}) = \mathrm{ch}_g(E', \nabla^{E'}) - \mathrm{ch}_g(E, \nabla^E). \tag{3.6}$$

3.3 Spectral sequence and the geometry

In this subsection, we analyze the geometry of the composition of fibrations to define a series of vector bundles over the base manifold S by Dirac operators. Furthermore, we define finite dimensional eta forms related to these vector bundles as the ingredients of remainder terms for family extension of adiabatic limit formula. We use the notation in subsection 3.1.

Let

$$D^H := \sum_{p=1}^{\dim Y} c(f_p) \nabla_{f_p^H}^{\mathscr{E}_Y \otimes \mathscr{E}_X}. \tag{3.7}$$

We define the $C^\infty(S)$-module

$$E_r := \{s_0 \in C^\infty(S, \mathscr{E}) : \text{There exist } s_0, \cdots, s_{r-1} \in C^\infty(S, \mathscr{E}),$$

$$\text{such that } D^X s_0 = 0,$$

$$D^H s_0 + D^X s_1 = 0, C s_0 + D^H s_1 + D^X s_2 = 0, \cdots,$$

$$C s_{r-3} + D^H s_{r-2} + D^X s_{r-1} = 0\}. \tag{3.8}$$

Here $b \in S$ and $C = D^Z - D^H - D^X$. We assume that for $r \geqslant 2$, E_r is a finite generated projective $C^\infty(S)$-module. Then it is the space of smooth sections of a vector bundle $\mathscr{E}_r$ over S. In [4], for Dolbeault operators, $\mathscr{E}_r$ are isomorphic to the spectral sequences. Moreover, on each $\mathscr{E}_r$, we define

$$D_r s_0 := p_r(D^H s_{r-1} + C s_{r-2}), \tag{3.9}$$

where $s_0 \in \mathscr{E}_r$ with s_{r-1}, s_{r-2} the elements in the definition (3.8). In [4] and [10], the Hodge theory can be used to draw the conclusion that

$$\ker D_r \simeq \mathscr{E}_{r+1} \tag{3.10}$$

for the flat case and the Dolbeault case. In [15], however, we show that this can be proved directly from the definitions (3.8) and (3.9) for general Dirac operators. On each $\mathscr{E}_r$, there are natural G-equivariant geometric triples $(\mathscr{E}_r, h_r, \nabla^r)$, defined by the projections from $(\mathscr{E}, \langle \cdot, \cdot \rangle_{\mathscr{E}}, \nabla^{\mathscr{E}})$. Note that $\dim \mathscr{E}_r$ is finite when $r \geqslant 2$. Then there exists r_0 such that $\mathscr{E}_r \cong \mathscr{E}_{r_0}$ whenever $r \geqslant r_0$. We denote by $\mathscr{E}_\infty$ the convergent one with geometric triple $(\mathscr{E}_\infty, h_\infty, \nabla^\infty)$. Furthermore, following the same assumption in Theorem 1 that $\dim \ker D_T^Z$ is constant for T large enough, we can show that $\mathscr{E}_\infty \simeq \ker D^Z$ (cf. [15, Lemma 3.6]).

With (3.10), we can define the finite dimensional eta forms:

$$\tilde{\eta}_g(\mathscr{E}_r, \mathscr{E}_{r+1}, \nabla^r, \nabla^{r+1}) := -\int_0^{+\infty} \tilde{\psi}_S \mathrm{Tr}_s \left[g \frac{\partial B_{r,t}}{\partial t} \exp(-B_{r,t}^2) \right], \quad r \geqslant 2, \tag{3.11}$$

where $B_{r,t} = \nabla^r + \sqrt{t} D_r$. With (3.11), the authors in [15] prove that the family extension of R can be written as

$$\tilde{R} = \sum_{r=2}^{r_0} \tilde{\eta}_g(\mathscr{E}_r, \mathscr{E}_{r+1}, \nabla^r, \nabla^{r+1}) + \tilde{\mathrm{ch}}_g(\ker D^Z, \nabla^\infty, \nabla^{\ker D^Z}). \tag{3.12}$$

3.4 Analysis with D_T^Z

In this subsection, we sketch the key idea to prove the result in [15], which heavily relies on the analytical localization technique developed in [4] and the estimates in [16].

As in [13, (2.64)], we rewrite the eta forms (2.10) and (3.5) as

$$\tilde{\eta}_g(\pi_{\underline{Z}}, \mathcal{E}) = -\int_0^{+\infty} \gamma^u(T, u) du,$$

$$\tilde{\eta}_g(\mathscr{E}_r, \mathscr{E}_{r+1}, \nabla^r, \nabla^{r+1}) = -\int_0^{+\infty} \gamma_r(u) du, \tag{3.13}$$

with $\gamma^u : [0, +\infty]_T \times [0, +\infty]_u \to \Omega^\bullet(S), \gamma_r : [0, +\infty]_u \to \Omega^\bullet(S), r = 2, 3, \cdots$
being constructed from Bismut superconnections:

$$\gamma^u(T, u) = \left\{ u^{-2} \psi_S \delta_{u^2} \widetilde{\mathrm{Tr}}[g \exp(-u^2 \mathcal{B}_T)] \right\}^{du},$$

$$\gamma_r(u) = \left\{ u^{-2} \psi_S \delta_{u^2} \widetilde{\mathrm{Tr}}[g \exp(-u^2 \mathcal{B}_r)] \right\}^{du}, \quad r \geqslant 2. \tag{3.14}$$

Here δ_{u^2} is the rescaling operator defined by $\delta_{u^2}(\alpha) = u^i \cdot \alpha$ for $\alpha \in \Lambda^i(T^*S) \widehat{\otimes} \mathscr{E}$,
and the definitions of $\mathcal{B}_T$ and $\mathcal{B}_r$ can be found in [15, (5.22)].

Moreover, it turns out that the spectrum of superconnections is the same
as those of the Dirac operators (cf. [15, Theorem 5.5]). Let $\mathbb{E}_b$ be the 0-order
Sobolev space of $C^\infty(Z_b, \pi_Z^* \Lambda(T^*S) \widehat{\otimes} \mathscr{E})$. Let U_0 be a region in the complex
plane, defined by the spectra of Dirac operators $D_r, r = 1, 2, \cdots$ (cf. [15, (5.7)]).
The key estimate can be obtained similarly as [4, Theorem 6.5].

Proposition 5   *Given $r \geqslant 2$, for $\lambda \in U_0, s \in \mathbb{E}$, there exists $C \in \mathbb{R}$, such that
when $T \to +\infty$,*

$$\|(\lambda - T^{r-1} D_T^Z)s - p_r(\lambda - D_r)^{-1} p_r s\| \leqslant \frac{C}{T} \|s\|. \tag{3.15}$$

This means that on the bundle $\mathscr{E}_r$, D_r can be estimated by $T^{r-1} D_T^Z$ when
$T \to +\infty$, which allows us to separate the spectrum of D_T^Z according to the order
of T in the asymptotic expansion to approximate D_r. Therefore, we define the
contours follows, where c_1, c_2 are defined to make all the nonzero eigenvalues of
$D_r, r = 2, 3, \cdots$ be enclosed by Δ_0.

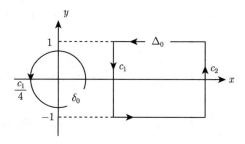

Figure 3.1   Contours δ_0, Δ_0

We shall use the functional calculus to express γ^u, γ_r in terms of the
spectrum of superconnections in (3.14):

$$F_{r,u,T} := u^{-2}\psi_S\delta_{u^2}\widetilde{\mathrm{Tr}}\left[\int_{\Delta_0} e^{-u^2\lambda}(\lambda - \mathcal{B}_{r,T})^{-1}d\lambda\right]. \quad r \geqslant 2;$$

$$F_{r,u,\infty} := u^{-2}\psi_S\delta_{u^2}\widetilde{\mathrm{Tr}}\left[\int_{\Delta_0} e^{-u^2\lambda}(\lambda - \mathcal{B}_r)^{-1}d\lambda\right], \quad r \geqslant 2; \tag{3.16}$$

$$G_{r,u,T} := u^{-2}\psi_S\delta_{u^2}\widetilde{\mathrm{Tr}}\left[\int_{\delta_0} e^{-u^2\lambda}(\lambda - \mathcal{B}_{r,T})^{-1}d\lambda\right], \quad r \geqslant 1;$$

$$G_{r,u,\infty} := u^{-2}\psi_S\delta_{u^2}\widetilde{\mathrm{Tr}}\left[\int_{\delta_0} e^{-u^2\lambda}(\lambda - \mathcal{B}_r)^{-1}d\lambda\right], \quad r \geqslant 1,$$

where $\mathcal{B}_{r,T}$ is the Bismut superconnection constructed by the Dirac operator $T^{r-1}D_T^Z$ due to (3.15). Then γ^u, γ_r can be written as, when $T \gg 1$

$$\gamma^u(T,u) = \{F_{1,u,T}\}^{du} + \{G_{1,u,T}\}^{du},$$
$$\gamma_r(u) = \{F_{1,u,\infty}\}^{du} + \{G_{1,u,\infty}\}^{du}, \quad r \geqslant 2. \tag{3.17}$$

By (3.17), (3.12) can be proved by the following lemma:

Lemma 6 ([15, Lemma 5.6]) *1. There exist $\delta, c, C, T_0 > 0$ such that for any $u \geqslant 1, T \geqslant T_0, r \geqslant 1$,*

$$|F_{r,u,T} - F_{r,u,\infty}| \leqslant \frac{C}{T^\delta}e^{-cu}, \quad |G_{r,u,T} - G_{r,u,\infty}| \leqslant \frac{C}{T^\delta}. \tag{3.18}$$

2. There exist $C, \delta > 0$, such that for any $u \in \mathbb{C}, |u| \leqslant 1, T \geqslant T_0$,

$$|F_{r,u,T} - F_{r,u,\infty}| \leqslant \frac{C}{T^\delta}, \quad |G_{r,u,T} - G_{r,u,\infty}| \leqslant \frac{C}{T^\delta}. \tag{3.19}$$

At the end, we show that the $\widetilde{R}$ in (3.12) indeed extends the remainder term R in (2.4) as we have promised. Let $g = e$ and S be a point in the definition of (3.11). We have that

$$\widetilde{\eta}_e(\mathscr{E}_r, \mathscr{E}_{r+1}, \nabla^r, \nabla^{r+1}) = \frac{1}{\sqrt{\pi}} \int_0^{+\infty} \mathrm{Tr}_s\left[\frac{D_r}{2\sqrt{t}}\exp(-tD_r^2)\right] dt. \tag{3.20}$$

By Gauss integration, $\dfrac{1}{\sqrt{\pi}}\displaystyle\int_0^{+\infty} \lambda e^{-\lambda^2 t}\dfrac{dt}{2\sqrt{t}} = \mathrm{sgn}(\lambda)$. Hence

$$\widetilde{\eta}_e(\mathscr{E}_r, \mathscr{E}_{r+1}, \nabla^r, \nabla^{r+1}) = \sum_{\lambda \in \mathrm{Spec}(D_r)} \mathrm{sgn}(\lambda). \tag{3.21}$$

By (3.15), we are able to make $\mathrm{Spec}(D_r)$ as $\{\lambda_T \in \mathrm{Spec}(D_T^Z) : \lambda_T \sim T^{-(r-1)}\}$.

References

[1] Atiyah M. F., Patodi V. K., and Singer I. M. *Spectral asymmetry and Riemannian geometry.* Bull. London Math. Soc. 5 (1973), 229–234.

[2] Atiyah M. F., and Singer I. M. *The index of elliptic operators: I.* Ann. Math. (1968), 484–530.

[3] Berline N., Getzler E., and Vergne M. *Heat Kernels and Dirac Operators.* New York: Springer-Verlag, 1992.

[4] Berthomieu A., and Bismut J.-M. *Quillen metrics and higher analytic torsion forms.* J. Reine Angew. Math. 457 (1994), 85–184.

[5] Bismut J.-M. *The Atiyah-Singer index theorem for families of Dirac operators: Two heat equation proofs.* Invent. Math. 83, 1 (1986), 91–151.

[6] Bismut J.-M., and Cheeger J. *η-invariants and their adiabatic limits.* J. Amer. Math. Soc. 2, 1 (1989), 33–70.

[7] Bismut J.-M., and Cheeger J. *Families index for manifolds with boundary, superconnections and cones. I, Families of manifolds with boundary and dirac operators.* J. Funct. Anal. 89, 2 (1990), 313–363.

[8] Bismut J.-M., and Cheeger J. *Families index for manifolds with boundary, superconnections and cones. II, The chern character.* J. Funct. Anal. 90, 2 (1990), 306–354.

[9] Bismut J.-M., and Freed D. S. *The analysis of elliptic families. I. Metrics and connections on determinant bundles.* Comm. Math. Phys. 106, 1 (1986), 159–176.

[10] Bunke U., and Ma X. *Index and secondary index theory for flat bundles with duality.* In: *Gil J., et al. (eds) Aspects of Boundary Problems in Analysis and Geometry.* Basel: Birkhäuser, 2004, 265–341.

[11] Dai X. *Adiabatic limits, nonmultiplicativity of signature, and Leray spectral sequence.* J. Amer. Math. Soc. 4, 2 (1991), 265–321.

[12] Donnelly H. *Eta invariants for g-spaces.* Indiana Univ. Math. J. 27, 6 (1978), 889–918.

[13] Liu B. *Functoriality of equivariant eta forms.* J. Noncommut. Geom. 11, 1 (2017), 225–307.

[14] Liu B. *Bismut-Cheeger eta form and higher spectral flow.* Int. Math. Res. Not. 2023, 13 (2022), 10964–10996.

[15] Liu B., and Zhan M. *Eta form and spectral sequence for the composition of fibrations.* arXiv: 2305.19637 (2023).

[16] Ma X. *Formes de torsion analytique et familles de submersions. II.* Asian J. Math. 4, 3 (2000), 633–668.

[17] Melrose R. B., and Piazza P. *Families of dirac operators, boundaries and the b-calculus.* J. Differential Geom. 46, 1 (1997), 99–180.

[18] Melrose R. B., and Piazza P. *An index theorem for families of dirac operators on odd-dimensional manifolds with boundary.* J. Differential Geom. 46, 2 (1997), 287–334.

[19] Quillen D. *Superconnections and the Chern character.* Topology 24, 1 (1985), 89–95.

[20] Witten E. *Global gravitational anomalies.* Comm. Math. Phys. 100, 2 (1985), 197–229.

Optimal Degeneration Problem of Fano Threefolds

MINGHAO MIAO

Department of Mathematics, Nanjing University, Nanjing, China

minghao.miao@smail.nju.edu.cn

Abstract This is a survey of determining the optimal degeneration of Fano threefolds. Some results of Kähler-Ricci flow, Kähler-Ricci soliton, and K-stability on Fano manifolds will be discussed.

1 Introduction

Finding canonical metric is a central problem in Kähler geometry. Suppose (M, ω) is a compact Kähler manifold. The Kähler metric ω is called a Kähler-Einstein (KE) metric if $\mathrm{Ric}(\omega) = \lambda\omega$ where $\lambda = -1, 0, 1$ by rescaling the metric. One necessary condition of X admitting Kähler-Einstein metric is that the first Chern class $c_1(M)$ has a definite sign, so we reduce to consider the following three cases: $c_1(M) < 0, c_1(M) = 0$ and $c_1(M) > 0$. For the case of $c_1(M) < 0$, it was proved by Aubin and Yau independently that M always admits Kähler-Einstein metric. And the case $c_1(M) = 0$ is a consequence of Yau's solution to Calabi conjecture, while it was known that there exists obstruction of the existence of Kähler-Einstein metric on $c_1(M) > 0$ case (i.e. for Fano manifolds). There is another type of canonical metric called Kähler-Ricci soliton (KRS) metric, which is a natural generalization of Kähler-Einstein metrics and is closely related to the limiting behavior of Kähler-Ricci flow. The Kähler metric ω is called the Kähler-Ricci soliton if it satisfies $\mathrm{Ric}(\omega) - \lambda\omega = L_\xi\omega$ where $L_\xi\omega$ is taking Lie derivative along the holomorphic vector field ξ. The case when $\lambda = 1, 0, -1$

corresponds to shrinking, steady, expanding Kähler-Ricci soliton. One fact is that steady and expanding Ricci soliton on compact manifold must be Einstein.

In 1997, Tian [Tia97] first introduced the concept of K-stability to characterize the existence of Kähler-Einstein metric on the Fano manifold. Later, it was proved in [Tia15] that a Fano manifold admits a Kähler-Einstein metric if and only if it is K-polystable. Other proofs were also given in [CDS15, DS16, CSW18, BBJ21, Zha21]. For the definition of K-stability, we will introduce the following equivalent version which is developed by Fujita-Li independently:

Theorem 1.1 ([Fuj19, Li17]) *Suppose (X, ω) is a Fano manifold, then*

(1) X is K-semistable if and only if for any prime divisor E over X, $\beta(E) \geqslant 0$.

(2) X is K-unstable if and only if for any prime divisor E over X, $\beta(E) < 0$.

Here a divisor E over X means there exists a proper birational morphism $\mu : Y \to X$ such that E is a prime divisor on Y, log discrepancy $A_X(E) := 1 + \mathrm{coeff}_E(K_Y - \mu^* K_X)$, expected vanishing order $S_X(E) = \dfrac{1}{\mathrm{Vol}(-K_X)} \cdot$
$\displaystyle\int_0^\infty \mathrm{Vol}(-\mu^* K_X - tE)\mathrm{d}t$ and $\beta(E) := A_X(E) - S_X(E)$.

Now, we turn to another aspect of this survey, which relates to Kähler-Ricci flow. Ricci flow was introduced by Richard Hamilton in [Ham82] in the 1980s which imitates the heat equation to deform the metric of a manifold to "average geometry". The evolution equation reads as follows: $\partial_t g_t = -2\mathrm{Ric}(g_t)$. After solving the Poincaré conjecture around the 2000s, geometers realized that Ricci flow is also a powerful tool in other geometric problems. In the world of Kähler geometry, Ricci flow will preserve Kählerian condition, so when restricted to a Kähler manifold, it is called the Kähler-Ricci flow. On a Fano manifold, we consider the following normalized Kähler-Ricci flow:

$$\begin{cases} \dfrac{\partial \omega(t)}{\partial t} = -\mathrm{Ric}\left(\omega(t)\right) + \omega(t) \\ \omega(0) = \omega_0 \end{cases} \tag{1.1}$$

where ω_0 and $\omega(t)$ denote the Kähler forms of initial Kähler metric g_0 and the solution $g(t)$ of Ricci flow, respectively. It was shown in [Cao85] that when ω_0 represents $2\pi c_1(X)$, equation (1.1) has a global solution $\omega(t)$ for all $t \geqslant 0$. Then it is natural to ask what is the limiting behavior of $\omega(t)$ as $t \to \infty$. It was proved in [TZ07, TZ13, TZZZ13, DS20] that if X admits a Kähler-Ricci soliton ω_{KRS} and ω_0 represents $2\pi c_1(X)$, then as t goes to ∞, the solution $\omega(t)$ of the normalized Kähler-Ricci flow (1.1) converges to ω_{KRS} up to the action of the automorphism group of X. It remains to see what we can expect if X does not admit any Kähler-Ricci soliton. In this situation, the complex structures will jump as $t \to \infty$ and it was conjectured in [Tia97], also referred to as Hamilton-Tian conjecture, that:

Conjecture 1.2 *For any global solution $\omega(t)$ of (1.1) as above, any sequence $(X, \omega(t))$ along Kähler-Ricci flow contains a subsequence converging to a $\mathbb{Q}$-Fano variety $(X_\infty, \omega_\infty)$ in the Gromov-Hausdorff topology, and $(X_\infty, \omega_\infty)$ admits a smooth shrinking Kähler-Ricci soliton outside the singular set S of X_∞ which is closed and of Hausdorff codimension at least 4. Moreover, this subsequence of $(X, \omega(t))$ converges locally to $(X_\infty, \omega_\infty)$ in the Cheeger-Gromov topology.*

This conjecture has been first solved for dimensions less than or equal to 3 ([TZ13]) and subsequently for higher dimensions ([Bam18, CW20, WZ04, BLXZ23]) by using different methods. The key to these methods is to establish a certain compactness for Kähler-Ricci flow which leads to the partial C^0-estimate along the flow. We would also like to mention that a generalized version of the Hamilton-Tian conjecture was proved in [Bam18] and an algebraic proof of the conjecture was given in [BLXZ23], building on the work of [HL23].

As we mentioned above, X_∞ coincides with X if and only if X admits Kähler-Ricci soliton. In this case, the solution of normalized Kähler-Ricci flow $\omega(t)$ converges to ω_{KRS} up to the action of the automorphism group of X. In general, the metric degeneration from X to X_∞ induces a finitely generated filtration on section ring $R = \oplus_m H^0(X, -mK_X)$. The metric degeneration has the following "two-step degeneration" description ([CSW18]). There is a special $\mathbb{R}$-test configuration $\mathcal{F}^{ss}$ with K-semistable central fiber (W, ξ) (we call it "semistable degeneration") and a $\mathbb{T}$-equivariantly special test configuration of (W, ξ) with K-polystable central fiber (X_∞, ξ) (we call it "polystable degeneration"). And X_∞ coincides with the Gromov-Hausdorff limit of the normalized Kähler-Ricci flow. It was shown in [DS20] that the semistable degeneration minimizes non-Archimedean $\mathbf{H}^{\mathrm{NA}}$-functional among all

$\mathbb{R}$-test configuration of X. Due to this minimizing picture, X_∞ is often called the "optimal degeneration" of X in the literature.

Here we can briefly summarize the relationship between the convergence of Kähler-Ricci flow and K-stability on the Fano manifold:

(1) If X is K-polystable, then X admits Kähler-Einstein metric, then $\omega(t) \to \omega_{KE}$ as $t \to \infty$ along normalized Kähler-Ricci flow.

(2) If X is K-semistable, then the Gromov-Hausdorff limit of the normalized Kähler-Ricci flow admits Kähler-Einstein metric.

(3) If X is K-unstable, then the Gromov-Hausdorff limit admits non-trivial Kähler-Ricci soliton.

We now turn to a brief discussion of the singularity formation of Kähler-Ricci flow. We recall that a solution $\omega(t)$ of Kähler-Ricci flow is of type I if the curvature of $\omega(t)$ is uniformly bounded, otherwise, we call $\omega(t)$ a solution of type II. There was a folklore conjecture closely related to Conjecture 1.2 stating that the Gromov-Hausdorff limit X_∞ is always smooth, i.e., the singular set S is always empty. This is equivalent to saying that the Kähler-Ricci flow has no type II solution. The folklore conjecture was first disproved in [LTZ18] by considering Fano compactifications of semisimple Lie groups. However, the lowest dimension among these group compactifications is 6. Naturally, one is led to wonder whether there exist examples of Fano manifold of lower dimensions on which Kähler-Ricci flow develops singularities of type II. Since the folklore conjecture holds for complex dimension two or less, a natural question is whether the lowest possibility is 3. We want to investigate Fano threefolds based on the recent progress of K-stability from the algebraic side.

2   Optimal Degeneration Problem of Fano Threefolds

Recently, the authors of [ACC+23] have been working on the Calabi problem of Fano threefolds, which aims to find all K-polystable smooth Fano threefolds. We can propose a more generalized version of the Calabi problem under the framework of the Hamilton-Tian theorem. If X is K-polystable, then the optimal degeneration is itself. If X is strictly K-semistable, then the optimal degeneration is the unique K-polystable degeneration [LWX21]. If X is K-unstable but admits Kähler-Ricci soliton, then the optimal degeneration is

X itself [TZ07, TZ13, TZZZ13, DS20]. If X is K-unstable but doesn't admit Kähler-Ricci soliton, then we need to determine the optimal degeneration X_∞ explicitly.

Due to the work of Iskovskikh-Mori-Mukai ([MM82, IP99]), we know every smooth Fano threefold belongs to one of the 105 deformation families. Based on the work of [ACC+23], we can list all deformation families of K-unstable Fano threefolds. There are 27 deformation families in total. 13 of them are toric manifolds. By [WZ04], they admit KRS. And 9 of them are T-variety of complexity one, [IS17, CS18] show that they all admit KRS. And 4 of them are T-variety of complexity two (No.2.26, No.2.28, No.3.14, No.3.16 in Mori-Mukai's list) and only one deformation family has discrete automorphism group (No.2.23). It is of most interest to determine the optimal degeneration in the remaining 5 deformation families (No. 2.23, No. 2.26, No. 2.28, No. 3.14, No. 3.16).

We prove the following result:

Theorem 2.1 ([MT22]) *Any Fano threefold X from family No.2.23 in Mori-Mukai's list has type II solutions of the normalized Kähler-Ricci flow. Namely, the Gromov-Hausdorff limit of X along the Kähler-Ricci flow is a singular $\mathbb{Q}$-Fano variety.*

In [Del22], Delcroix constructed examples of K-unstable Fano manifolds by blowing up quadrics along lower dimensional linear subquadrics and showed that they do not admit any Kähler-Ricci soliton. The lowest dimension of these examples of Delcroix is 5, so he further asked for examples of Fano threefolds or fourfolds which are K-unstable and do not admit Kähler-Ricci soliton. The Theorem 2.1 gives an affirmative answer to Question 1.3 in [Del22].

Question 2.2([Del22, Question 1.3]) *Do there exist examples of Fano threefolds with no Kähler-Ricci solitons that are K-unstable? What about Fano fourfolds?*

Theorem 2.3 ([MT22]) *Any Fano threefold X from the family No.2.23 in the Mukai-Mori's list does not admit Kähler-Ricci soliton and is K-unstable.*

Here we briefly mention the strategy of the proof of the main theorem of [MT22].

(1) Show automorphism group of X is finite. Therefore, there is no non-

trivial holomorphic vector field.

(2) There exists a prime divisor E on X such that beta-invariant $\beta_X(E) < 0$. Hence, by Fujita-Li's valuative criterion, X is K-unstable. So X doesn't admit KE.

(3) No non-trivial holomorphic vector field and no KE will imply no KRS.

(4) Prove by contradiction if we have smooth deformation $\pi : \mathcal{X} \to \Delta$, then use the classification theory of smooth Fano threefold to conclude Gromov-Hausdorff limit still lies in No.2.23, which contradicts with Gromov-Hausdorff limit admits KRS.

Inspired by the optimal degeneration problem of Fano threefolds, we can propose the following questions:

Question 2.4   *Suppose X is a Fano threefold from the family No.2.23, can we find the Gromov-Hausdorff limit X_∞ explicitly? Are members from the family No.2.23 the only examples of Fano threefolds on which Kähler-Ricci flow has solutions of type II?*

Remark 2.5   *Recently, [MW23] develops the weighted Abban-Zhuang estimate and proves the existence of Kähler-Ricci soliton on Fano threefolds from No. 2.28 and No.3.14 in the Mori-Mukai's list. Thus, the optimal degeneration of members from these two deformation families coincides with itself.*

References

[ACC⁺23]   Carolina Araujo, Ana-Maria Castravet, Ivan Cheltsov, et al. *The Calabi Problem for Fano Threefolds.* Cambridge: Cambridge University Press, 2023.

[Bam18]   Richard Bamler. *Convergence of Ricci flows with bounded scalar curvature.* Ann. Math., 188(3): 753–831, 2018.

[BBJ21]   Robert Berman, Sebastian Boucksom, and Mattias Jonsson. *A variational approach to the Yau-Tian-Donaldson conjecture.* J. Amer. Math. Soc., 34(3): 605–652, 2021.

[BLXZ23]   Harold Blum, Yuchen Liu, Chenyang Xu, et al. *The existence of the Kähler-Ricci soliton degeneration.* Forum Math. Pi, 11: Paper No. e9, 28, 2023.

[Cao85]   Huai-Dong Cao. *Deformation of Kähler metrics to Kähler-Einstein metrics on compact Kähler manifolds.* Invent. Math., 81(2): 359–372, 1985.

[CDS15]   Xiuxiong Chen, Simon Donaldson, and Song Sun. *Kähler-Einstein metrics on Fano manifolds. I: Approximation of metrics with cone singularities, II:*

Limits with cone angle less than 2π, *III: Limits as cone angle approaches* 2π *and completion of the main proof.* J. Amer. Math. Soc., 28(1): 183–197, 199–234, 235–278, 2015.

[CS18]   Jacob Cable and Hendrik Süß. *On the classification of Kähler-Ricci solitons on Gorenstein del Pezzo surfaces.* Eur. J. Math., 4(1): 137–161, 2018.

[CSW18]   Xiuxiong Chen, Song Sun, and Bing Wang. *Kähler-Ricci flow, Kähler-Einstein metric, and K-stability.* Geom. Topol., 22(6): 3145–3173, 2018.

[CW20]   Xiuxiong Chen and Bing Wang. *Space of Ricci flows (II)—Part B: Weak compactness of the flows.* J. Differential Geom., 116(1): 1–123, 2020.

[Del22]   Thibaut Delcroix. *Examples of K-unstable Fano manifolds.* Ann. Inst. Fourier (Grenoble), 72(5): 2079–2108, 2022.

[DS16]   Ved Datar and Gábor Székelyhidi. *Kähler-Einstein metrics along the smooth continuity method.* Geom. Funct. Anal., 26(4): 975–1010, 2016.

[DS20]   Ruadhaí Dervan and Gábor Székelyhidi. *The Kähler-Ricci flow and optimal degenerations.* J. Differ. Geom., 116(1): 187–203, 2020.

[Fuj19]   Kento Fujita. *A valuative criterion for uniform K-stability of* $\mathbb{Q}$-*Fano varieties.* J. Reine Angew. Math., 751: 309–338, 2019.

[Ham82]   Richard S. Hamilton. *Three-manifolds with positive Ricci curvature.* J. Differ. Geom., 17(2): 255–306, 1982.

[HL23]   Jiyuan Han and Chi Li. *On the Yau-Tian-Donaldson conjecture for generalized Kähler-Ricci soliton equations.* Comm. Pure Appl. Math., 76(9): 1793–1867, 2023.

[IP99]   V. A. Iskovskikh and Yu. G. Prokhorov. *Fano varieties//Algebraic Geometry, V*, volume 47 of *Encyclopaedia Math. Sci.*, pages 1–247. Berlin: Springer, 1999.

[IS17]   Nathan Ilten and Hendrik Süß. *K-stability for Fano manifolds with torus action of complexity 1.* Duke Math. J., 166(1): 177–204, 2017.

[Li17]   Chi Li. *K-semistability is equivariant volume minimization.* Duke Math. J., 166(16): 3147–3218, 2017.

[LTZ18]   Yan Li, Gang Tian, and Xiaohua Zhu. *Singular limits of Kähler-Ricci flow on Fano G-manifolds.* (to appear in *Amer. J. Math.*), arXiv:1807.09167, 2018.

[LWX21]   Chi Li, Xiaowei Wang, and Chenyang Xu. *Algebraicity of the metric tangent cones and equivariant K-stability.* J. Amer. Math. Soc., 34(4): 1175–1214, 2021.

[MM82]   Shigefumi Mori and Shigeru Mukai. *Classification of Fano 3-folds with* $B_2 \geqslant 2$. Manuscripta Math., 36(2): 147–162, 1981/82.

[MT22]   Minghao Miao and Gang Tian. *A note on Kähler-Ricci flow on Fano threefolds.* (accepted by *Peking Math. J.*), 2022. arXiv:2210.15263.

[MW23]   Minghao Miao and Linsheng Wang. *Kähler-Ricci solitons on Fano threefolds with non-trivial moduli.* 2023. arXiv:2309.14212.

[Tia97]   Gang Tian. *Kähler-Einstein metrics with positive scalar curvature.* Invent. Math., 130(1): 1–37, 1997.

[Tia15]   Gang Tian. *K-stability and Kähler-Einstein metrics.* Comm. Pure Appl. Math., 68(7): 1085–1156, 2015.

[TZ07]   Gang Tian and Xiaohua Zhu. *Convergence of Kähler-Ricci flow.* J. Amer. Math. Soc., 20(3): 675–699, 2007.

[TZ13]   Gang Tian and Xiaohua Zhu. *Convergence of the Kähler-Ricci flow on Fano manifolds.* J. Reine Angew. Math., 678: 223–245, 2013.

[TZZZ13] Gang Tian, Shijin Zhang, Zhenlei Zhang, and Xiaohua Zhu. *Perelman's entropy and Kähler-Ricci flow on a Fano manifold.* Trans. Amer. Math. Soc., 365(12): 6669–6695, 2013.

[WZ04]   Xu-Jia Wang and Xiaohua Zhu. *Kähler-Ricci solitons on toric manifolds with positive first Chern class.* Adv. Math., 188(1): 87–103, 2004.

[Zha21]   Kewei Zhang. *A quantization proof of the uniform Yau-Tian-Donaldson conjecture.* (accepted by *JEMS*), arXiv:2102.02438, 2021.

Spaces with Ricci Curvature Lower Bounds

JIAYIN PAN

Department of Mathematics, University of California, Santa Cruz, USA

E-mail: jpan53@ucsc.edu

GUOFANG WEI

Department of Mathematics, University of California, Santa Barbara, USA

E-mail: wei@ucsb.edu

Abstract We give a survey on authors' recent works on the geometry and topology of manifolds with Ricci curvature lower bounds or Ricci limit spaces. The topics include the Busemann functions and fundamental groups on complete manifolds with nonnegative Ricci curvature, as well as the Hausdorff dimension and local topology of Ricci limit spaces.

1 Busemann function of manifolds with nonnegative Ricci curvature

1.1 Busemann function

Let M^n be a complete noncompact manifold. For any $p \in M$, there exists a unit speed ray $\gamma(t) : [0, +\infty) \to M$, i.e. $d(\gamma(t), \gamma(s)) = |t - s|$ for all $t, s \in [0, \infty)$. The Busemann function associated with γ is a renormalized distance function to infinity along γ, which plays an important in the study of noncompact manifolds.

J. Y. Pan is supported partially by NSF DMS 2304698 and Simons Foundation Travel Support for Mathematicians. G. F. Wei is supported partially by NSF DMS grant 2104704.

Definition 1   The Busemann function associated to a ray γ is a function $b_\gamma : M \to \mathbb{R}$ defined by

$$b_\gamma(x) = \lim_{t \to \infty} \Big(t - \mathrm{d}(x, \gamma(t)) \Big).$$

Note that the sequence is monotone and bounded so the limit exists.

Remark 2   • b_γ is Lipschitz with Lipschitz constant 1.
• Along γ, $b_\gamma(\gamma(t)) = t$ is linear in t.

Example 3   Let $M = \mathbb{R}^n$ with the usual Euclidean metric. Then all rays are of the form $\gamma(t) = \gamma(0) + t\gamma'(t)$, and $b_\gamma(x) = \langle x - \gamma(0), \gamma'(0) \rangle$.

Theorem 4 (Cheeger-Gromoll, 71', 72')
• [8] If the sectional curvature $K_M \geqslant 0$, then $\mathrm{Hess}\, b_\gamma \geqslant 0$.
• [7] If the Ricci curvature $\mathrm{Ric}_M \geqslant 0$, then $\Delta b_\gamma \geqslant 0$.
   (both in barrier sense)

Remark 5   1. The first result plays an important role in the proof of Soul's theorem, while the second one leads to the splitting theorem.

2. By Laplacian comparison, we have $\Delta r \leqslant \Delta_{\mathbb{R}^n} \bar{r} = \dfrac{n-1}{r}$. So intuitively,

$$\Delta \Big(t - \mathrm{d}(x, \gamma(t)) \Big) \geqslant -\frac{n-1}{\mathrm{d}(x, \gamma(t))} \to 0 \text{ as } t \to \infty.$$

Definition 6 (Busemann function of a point) $b_p(x) := \sup_\gamma b_\gamma(x)$, where the supremum is taken among all rays γ starting from p.

When M^n is polar with a pole at p, then $b_p(x) = \mathrm{d}(p, x)$. Still $b_p(x)$ is convex when $K_M \geqslant 0$, and subharmonic when $\mathrm{Ric}_M \geqslant 0$ in the barrier sense.

The convexity of b_p implies $b_p(x)$ is proper. In fact, it implies that the sublevel set $b_p^{-1}(-\infty, a] = \cap_\gamma b_\gamma^{-1}(-\infty, a]$ is compact.

Question 7 (Open problem since 70's)   Is b_p proper when $\mathrm{Ric}_M \geqslant 0$?

It has been shown that the answer is yes in many special cases:
• When M is polar with a pole at p, we have $b_p(x) = \mathrm{d}(p, x)$, proper.
• If $\displaystyle\limsup_{r \to \infty} \frac{\mathrm{diam}\,(\partial B(p, r))}{r} = \epsilon < 1$, then $\displaystyle\liminf_{x \to \infty} \frac{b_p(x)}{\mathrm{d}(p, x)} \geqslant 1 - \epsilon > 0$, which implies b_p is proper.

- **Shen, 1996** ([25]): When M^n has Euclidean volume growth, b_p is proper.
- **Sormani 1998** ([26]): When M^n has linear volume growth, i.e., $Cr \leqslant \mathrm{vol}B(x, r) \leqslant C'r$, then b_p is proper.

In Section 1.3, we will see that the answer in general is negative.

1.2 Nabonnand's example of manifolds with positive Ricci curvature

We first recall Nabonnand's example [15], which is the first example of a manifold with positive Ricci curvature and infinite fundamental group π_1.

Example 8 Let $M = \mathbb{R}^k \times \mathbb{S}^1$ equipped with the doubly warped metric

$$[0, \infty) \times_f S^{k-1} \times_h S^1, \quad g = \mathrm{d}r^2 + f^2(r) \, \mathrm{d}s_{k-1}^2 + h^2(r) \, \mathrm{d}s_1^2$$

with $f(0) = 0, f'(0) = 1, f''(0) = 0, h(0) > 0, h'(0) = 0$. Denote $H = \dfrac{\partial}{\partial r}$, u a unit vector tangent to $\mathbb{S}^{k-1}$, and v a unit vector tangent to $\mathbb{S}^1$. Then one can compute

$$\mathrm{Ric}(H, H) = -(k-1)\frac{f''}{f} - \frac{h''}{h} \tag{1.1}$$

$$\mathrm{Ric}(u, u) = -\frac{f''}{f} - \frac{k-2}{f^2}\left(1 - (f')^2\right) - \frac{f'h'}{fh} \tag{1.2}$$

$$\mathrm{Ric}(v, v) = -\frac{h''}{h} - (k-1)\frac{f'h'}{fh}. \tag{1.3}$$

When $0 < f' < 1, f'' < 0, h' < 0$, and $k \geqslant 2$, it is easy to see that $\mathrm{Ric}(u, u) > 0$.

Choose $h = f'$, then $\mathrm{Ric}(v, v) = \mathrm{Ric}(H, H)$. Let f be the solution of the ODE:

$$\begin{cases} f' = (1 - \varphi(f))^{\frac{1}{2}} \\ f(0) = 0, \end{cases}$$

where $\varphi(x) = \dfrac{\sqrt{3}}{\pi} \displaystyle\int_0^x \frac{\arctan u^3}{u^2} \, \mathrm{d}u$. Here we choose an explicit φ, there are many other choices of φ for the construction. As $\displaystyle\int_0^\infty \frac{\arctan u^3}{u^2} \, \mathrm{d}u = \frac{\pi}{\sqrt{3}}$, we have $0 < \varphi(x) < 1$ for $x \in (0, \infty)$. Note that $h \to 0$ and $h \sim r^{-1/3}$, $f \sim r^{2/3}$ as $r \to +\infty$.

Then one computes that $\mathrm{Ric}(H, H) > 0$ when $k \geqslant 3$.

Same construction works for $\mathbb{R}^k \times M^q$, where M has nonnegative Ricci curvature by modifying with $h = (f')^{1/q}$ [1].

Example 9 In [33] the second-named author constructed a metric with positive Ricci curvature on $\mathbb{R}^k \times N$, where N is a nilmanifold and k is large, as warped products

$$[0, \infty) \times_f S^{k-1} \times (N, g_r), \quad g = dr^2 + f^2(r) \, ds_{k-1}^2 + g_r.$$

This is the first example of manifolds with positive Ricci curvature with nilpotent fundamental group. See [2,3] for more constructions along the line.

1.3 Example of manifolds with positive Ricci curvature and non-proper Busemann function

Theorem 10 ([22]) *Given any integer $n \geqslant 4$, there is an open n-manifold with positive Ricci curvature and a non-proper Busemann function.*

Proof First we study the geodesics in Nabonnand's example. Note that for each $x \in S^{k-1}$, the subset

$$C(x) = \{(r, \pm x, v) | r \geqslant 0, v \in S^1\}$$

is a totally geodesic and geodesically complete submanifold in M. Given any fixed point p at $r = 0$ in M, there are three types of geodesics starting at p:

(i) Moving purely in $\mathbb{R}^k$, that is, $\gamma(t) = (t, x, y)$, where $x \in S^{k-1}$ and $y \in S^1$ are independent of t.

(ii) Moving purely in the S^1 direction, that is, $\gamma(t) = (0, x, c(t))$, where $x \in S^{k-1}$ and $c(t)$ goes around the circle.

(iii) A mixture of both, that is, $\gamma(t) = (r(t), x, y(t))$, where $x \in S^{k-1}$.

In case (i), the geodesic is a ray and in case (ii), the geodesic is a closed circle. In case (iii), by Clairaut's relation $h(r(t)) \cdot (\cos \theta(t)) = \text{const} = \cos \theta(0)$, where $\theta(t)$ is the angle between γ and the parallel circle at $\gamma(t)$. Since h goes to zero as r increases and $\cos \theta$ is bounded, the geodesic stays in a bounded region and crosses $\{r = 0\}$ transversely infinite many times.

On the universal cover $\tilde{M}$ of M, a geodesic $\tilde{\gamma}$ starting at $\tilde{p}$ is a ray precisely when the projection $\pi(\tilde{\gamma})$ is case (i). This is because in cases (ii) and (iii), the geodesics are bounded in M, then they cannot lift to rays. Otherwise, the π_1-action would give a line in $\tilde{M}$, contradicting $\text{Ric} > 0$.

Let g be a generator of $\pi_1(M, p) = \mathbb{Z}$. We claim $b_{\tilde{p}}(g^l\tilde{p}) = 0$ for all $l \in \mathbb{Z}$. Since the orbit at $\tilde{p}$ is noncompact as $\mathbb{Z} = \langle g \rangle$ is an infinite group, $b_{\tilde{p}}$ is not proper.

To show the claim, we prove that $b_{\tilde{\gamma}}(g^l\tilde{p}) = 0$ for any ray $\tilde{\gamma}$ at $\tilde{p}$. Note that we have shown that $\tilde{\gamma}$ must be the lift of a geodesic described in case (i). Now

$$b_{\tilde{\gamma}}(g^l\tilde{p}) = \lim_{t\to\infty} (t - d(\tilde{\gamma}(t), g^l\tilde{p})).$$

Let $\tilde{\alpha}$ be a minimal geodesic connecting $\tilde{\gamma}(t)$ and $g^l\tilde{p}$, then

$$d(\tilde{\gamma}(t),\ g^l\tilde{p}) = l(\tilde{\alpha}) = \text{length of the shortest curve in the homotopy class}$$

$$\leqslant t + l \cdot 2\pi h(t).$$

Covering map is distance non-increasing, hence

$$d(\tilde{\gamma}(t),\ g^l\tilde{p}) \geqslant d(\gamma(t), p) = t.$$

Therefore

$$-2\pi l \cdot h(t) \leqslant t - d(\tilde{\gamma}(t), g^l\tilde{p}) \leqslant 0.$$

Since $h(t) \to 0$ as $t \to \infty$, we have $b_{\tilde{\gamma}}(g^l\tilde{p}) = 0$ and proved the claim. □

Question 11   *What about $n = 3$?*

Bruè-Naber-Semola [4] recently gave examples of manifolds M^n with nonnegative Ricci curvature and π_1 not finitely generated for $n \geqslant 7$.

Question 12   *Is the properness of the Busemann function related to the finite generation of the π_1?*

2   Hausdorff dimension of Ricci limit spaces

2.1   Hausdorff dimension

Recall that the Hausdorff measure is an outer measure on subsets of a general metric space (X, d).

Definition 13 (Hausdorff Measure)   Let $0 \leqslant \alpha < \infty$ and $0 < \delta < \infty$. Let $A \subset X$. Define an outer measure

$$\mathcal{H}^\alpha_\delta(A) := \omega_\alpha \cdot \inf\left\{\sum_{i=1}^\infty \left(\tfrac{\text{diam}\, C_i}{2}\right)^\alpha \,\middle|\, A \subset \bigcup_{i=1}^\infty C_i, \text{ with } \text{diam}\, C_i \leqslant \delta \text{ for all } i \in \mathbb{N}\right\},$$

the infimum is taken over all countable covers of A.

The α-dimensional *Hausdorff measure* of $A \subset X$ is the outer measure

$$\mathcal{H}^\alpha(A) := \lim_{\delta \to 0} \mathcal{H}^\alpha_\delta(A).$$

Definition 14 The quantity

$$\dim_{\mathcal{H}}(A) := \inf\{0 \leqslant s < \infty \mid \mathcal{H}^s(A) = 0\} = \inf\{0 \leqslant s < \infty \mid \mathcal{H}^s(A) \neq \infty\}$$

is called the *Hausdorff dimension* of A.

Example 15 Let $\alpha \geqslant 1$ and let $(\mathbb{R}, d_\alpha)$ be defined by

$$d_\alpha(t_1, t_2) = |t_1 - t_2|^{1/\alpha}.$$

One can see that $\dim_{\mathcal{H}}(\mathbb{R}, d_\alpha) = \alpha$.

If $(M_i^n, p_i, \mu_i) \overset{mGH}{\to} (Y, p, \mu)$ and each M_i has Ric $\geqslant -(n-1)H$, where $\mu_i = \mathrm{vol}(\cdot)/\mathrm{vol}(B(p_i, 1))$ is the renormalized measure on M_i, then under the limit renormalized limit measure μ we have $r_1 \geqslant r_2$ implying

$$\frac{\mu(B(y, r_1))}{\mu(B(y, r_2))} \geqslant \frac{v(n, H, r_1)}{v(n, H, r_2)}, \tag{2.1}$$

where $v(n, H, r)$ means the volume of an r-ball in the n-dimensional space form with constant curvature H. This relative volume comparison implies the following.

Proposition 16 ([6]) *Any such space Y has $\dim_{\mathcal{H}}(Y) \leqslant n$.*

Let (M_i^n, g_i, p_i) be a sequence of Riemannian manifolds with $\mathrm{Ric}_{M_i} \geqslant (n-1)$ H and $M_i^n \overset{GH}{\to} Y$. Recall that Y is called a non-collapsed limit if

$$\mathrm{vol}(B(p_i, 1)) \geqslant v > 0,$$

and collapsed if

$$\mathrm{vol}(B(p_i, 1)) \to 0.$$

Let μ be a renormalized measure on Y. We have the following result

Proposition 17 ([6]) *If Y is a non-collapsed limit, then $\mu = c\mathcal{H}^n$ for some constant $c > 0$ and $\dim_{\mathcal{H}}(Y) = n$.*

If Y is a collapsed limit, then $\dim_{\mathcal{H}}(Y) \leqslant n-1$. In particular, the Hausdorff dimension of Y cannot be between n and $n-1$.

2.2 Rectifiable dimension of Ricci limit spaces

Definition 18 (tangent cone at a point) Let (X, d) be a metric space. Let $p \in X$. We take the pGH limit of $(X, p, \lambda_n d)$ where $\lambda_n \to \infty$. If such a limit exists, then it is called a tangent cone of X at p.

Remark 19   *Note that the limit may depend on the choice of sequence λ_n, i.e. tangent cones may not be unique. If it is unique, we denote the tangent cone at p as $C_p(X)$. The intuition is that we are zooming in the space at p.*

Definition 20 (asymptotic cone) Let (X, d) be a (noncompact) metric space. Let $p \in X$. We take the pGH limit of $(X, p, \lambda_n d)$ where $\lambda_n \to 0$. If such a limit exists, then it is called an asymptotic cone of X.

Remark 21   *The asymptotic cone does not depend on $p \in X$, but may depend on λ_n. The intuition is that we are zooming out the space, looking from somewhere far away. The asymptotic cone is especially useful when studying spaces with* Ric $\geqslant 0$.

Definition 22 (regular and singular points)   A point $p \in X$ is called a regular point if the tangent cone at p exists, and is unique and isometric to $\mathbb{R}^k$ for some integer k. We denote $\mathcal{R} = \{p \in X | p \text{ is regular}\}$, the collection of all regular points in X, $\mathcal{S} = X \setminus \mathcal{R}$, the set of singular points.

Remark 23   *In general, different regular points in X may have different k. For example, consider a suitable CW complex.*

Denote the k-regular set,

$$\mathcal{R}_k = \{y \in Y \mid C_y(Y) \text{ is unique and isometric to } \mathbb{R}^k\}.$$

We have the following general result for Ricci limit space Y by Colding-Naber.

Theorem 24 ([9])   *There exists a unique integer k, $0 \leqslant k \leqslant n$ such that $\mathcal{R}_k$ has full μ-measure for any limit renormalized measure μ, i.e., $\mu(Y \setminus \mathcal{R}_k) = 0$.*

This k is called the rectifiable dimension or essential dimension. In general, k is not equal to the Hausdorff dimension of Y, see Section 2.3 below. However, in the non-collapsed case, we indeed have [6]

$$k = \dim_{\mathcal{H}}(Y) = \dim(M_i).$$

In general, $\dim_{\mathcal{H}} \mathcal{R}_k = k$.

2.3 Examples of Ricci limit space with Hausdorff dimension different from the rectifiable dimension

Consider the doubly warped product

$$M = [0, \infty) \times_f S^{k-1} \times_h S^1, \quad g = dr^2 + f^2(r)\, ds_{k-1}^2 + h^2(r)\, ds_1^2$$

again as in Section 1.2, but with warping functions as in [33]. Namely, let $f(r) = r(1+r^2)^{-\frac{1}{4}} \sim \sqrt{r}$ and $h(r) = (1+r^2)^{-\alpha} \sim r^{-2\alpha}$, where $\alpha > 0$.

From the curvature formulas (1.1)–(1.3) one can check that Ric > 0 if $k \geqslant \max\{4\alpha + 3, 16\alpha^2 + 8\alpha + 1\}$.

Let $\tilde{M}$ be its universal cover, which is diffeomorphic to $\mathbb{R}^{k+1}$. We denote the asymptotic cone, singular set, regular set of $\tilde{M}$ by $Y, \mathcal{S}, \mathcal{R}_2$, respectively.

Theorem 25 ([21])
> 1. $Y = [0, +\infty) \times \mathbb{R}, \ \mathcal{S} = \{0\} \times \mathbb{R}, \ \mathcal{R}_2 = (0, +\infty) \times \mathbb{R}$
> 2. $\dim_{\mathcal{H}}(\mathcal{S}) = 1 + 2\alpha, \ \dim_{\mathcal{H}}(\mathcal{R}_2) = 2$

Theorem 26 ([10]) *Y is the metric completion of an incomplete Riemannian metric $g_Y = dr^2 + r^{-4\alpha}\, dv^2$ on $(0, +\infty) \times \mathbb{R}$.*

Remark 27 *Y with this metric is called 2α-Grushin halfplane, an almost Riemannian and RCD space at the same time. We remark that the whole Grushin plane is not an RCD space as the singular set cuts the plane into two disjoint parts but the regular set of RCD space should be path connected.*

Theorem 26 has some immediate interesting consequences.

For $\lambda > 0$, consider $F_\lambda(r, v) = (\lambda r, \lambda^{1+2\alpha}v)$. Then we have $F_* g_Y = \lambda^2 g_Y$, therefore $d\left(F_\lambda(y_1), F_\lambda(y_2)\right) = \lambda d(y_1, y_2)$. In other words, $\{F_\lambda\}$ are metric dilations of Y. Apply above with $\lambda = v^{\frac{1}{1+2\alpha}}$. Then

$$d\left((0, v), (0, 0)\right) = d\left(F_\lambda(0, 1), F_\lambda(0, 0)\right) = v^{\frac{1}{1+2\alpha}} d\left((0, 1), (0, 0)\right).$$

This implies $\dim_{\mathcal{H}}(\mathcal{S}) = 1 + 2\alpha$ from Example 15.

Proof of Theorem 26 We have $\tilde{M} = \mathbb{R}^k \times \mathbb{R}^1$, with doubly warped metric

$$g = dr^2 + r^2(1+r^2)^{-\frac{1}{2}} ds_{k-1}^2 + (1+r^2)^{-2\alpha} dv^2.$$

Given any $\lambda > 1$, let $s = \lambda^{-1}r$, $w = \lambda^{-2\alpha}v$. Then we get

$$\lambda^{-2}g_{\tilde{M}} = \lambda^{-2}\left[dr^2 + r^2(1+r^2)^{-\frac{1}{2}}ds_{k-1}^2 + (1+r^2)^{-2\alpha}dv^2\right]$$

$$= ds^2 + \frac{s^2}{1+\lambda^2 s^2}ds_{k-1}^2 + (1+\lambda^2 s^2)^{-2\alpha}\lambda^{4\alpha}dw^2$$

As we take the limit $\lambda \to \infty$, this metric approaches $ds^2 + s^{-4\alpha}dw^2$. $\qquad\square$

Question 28   *Given $\alpha > 1$, what is the smallest dimension n such that a Ricci limit space $X \in \mathcal{M}(n, -1)$ admits an isometric $\mathbb{R}$-orbit with Hausdorff dimension α?*

3   Characterizing virtual abelianness/nilpotency of $\pi_1(M)$

3.1   Small escape rate and virtual abelianness

We have mentioned in Section 1.2 that $\mathbb{R}^k \times N$, where N is a nilmanifold, admits a metric with positive Ricci curvature when k is large. This is distinct from open manifolds with nonnegative sectional curvature, whose fundamental groups are virtually abelian (i.e., contains an abelian subgroup of finite index). In fact, if M has $\sec_M \geqslant 0$, then it follows from Cheeger-Gromoll soul theorem that M is homotopic to a closed submanifold S with $\sec_S \geqslant 0$ in M, thus $\pi_1(M) = \pi_1(S)$ is virtually abelian.

Therefore, it is natural to investigate on what additional conditions $\pi_1(M)$ is virtually abelian for nonnegative Ricci curvature; or equivalently, we can ask how virtual abelianness or nilpotency of $\pi_1(M)$ is related to the geometry of M.

It follows from Cheeger-Gromoll splitting theorem that we have the following result on virtual abelianness.

Proposition 29   *Let (M, x) be an open manifold of $\mathrm{Ric} \geqslant 0$. If $\sup\limits_{\gamma \in \Gamma} d_H(x, c_\gamma) < \infty$, where c_γ is a minimal representing loop of $\gamma \in \pi_1(M, x)$ at x, then $\pi_1(M, x)$ is virtually abelian.*

The assumption in Proposition 29 always holds when $\sec \geqslant 0$. This follows from Cheeger-Gromoll soul theorem and Sharafutdinov retraction [8,24]. On the other hand, the assumption in Proposition 29 is quite restrictive for manifolds with nonnegative Ricci curvature. In fact, if M has positive Ricci curvature and an infinite fundamental group, then it follows from the Cheeger-Gromoll splitting theorem that the representing geodesic loops c_γ will always escape from any bounded sets as γ exhausts $\pi_1(M, x)$.

To study this escape phenomenon and its relation to $\pi_1(M)$, the first-named author introduced the notion of escape rate in [16].

Definition 30 Let (M, x) be an open manifold with an infinite fundamental group. We define the *escape rate* of (M, x), a scaling invariant, as

$$E(M, x) = \limsup_{|\gamma| \to \infty} \frac{d_H(x, c_\gamma)}{|\gamma|},$$

where $\gamma \in \pi_1(M, x)$, $|\gamma| = \text{length}(c_\gamma)$, and d_H is the Hausdorff distance. If $\pi_1(M)$ is finite, then we set $E(M, x) = 0$ as a convention.

By definition, the escape rate takes value within $[0, 1/2]$. It is known that $E(M, p) < 1/2$ implies the finite generation of $\pi_1(M)$ by Sormani's halfway lemma [27]. It is unclear whether its converse holds.

Question 31 *Let (M, p) be an open n-manifold with $\text{Ric} \geq 0$ and a finitely generated fundamental group. Is it true that $E(M, p) < 1/2$?*

Regarding examples as doubly warped products $[0, \infty) \times_f S^{p-1} \times_h S^1$, their escape rates are determined by the warping function h. As $h(r)$ decreases to 0 as $r \to \infty$, the representing geodesic loop will take advantage of the thin end to shorten its length, which will in turn enlarge its size. In other words, one should expect that the faster h decays, the larger the escape rate is. In fact, if $h(r)$ has polynomial decay, then $E(M, x)$ is positive; if $h(r)$ has logarithm decay or converges to a positive constant, then $E(M, x) = 0$.

As the main result in [17], small escape rate implies virtual abelianness. Note that this greatly generalized Proposition 29.

Theorem 32([17]) *Given n, there exists a constant $\epsilon(n) > 0$ such that if (M, x) is an open n-manifold with $\text{Ric} \geq 0$ and $E(M, x) \leq \epsilon(n)$, then $\pi_1(M, x)$ is virtually abelian.*

In other words, if $\pi_1(M, x)$ contains a torsion-free nilpotent subgroup of nilpotency step ≥ 2, then its minimal representing loops must escape any bounded subsets at a relatively fast rate compared to their lengths.

The proof of Theorem 32 depends on the equivariant asymptotic geometry of $(\widetilde{M}, \pi_1(M, x))$. The key is to show that any asymptotic π_1-orbit at $\tilde{x}$ is Gromov-Hausdorff close to a Euclidean factor. See [16, 17] for details.

3.2 Nilpotency step and Hausdorff dimension

Given Theorem 32, it is natural to further study the equivariant asymptotic geometry without the smallness of escape rate.

To better motivate the next result, we recall the structure of Carnot groups [23]. Let Γ be a finitely generated virtually nilpotent group with nilpotency step l. Any finite generating set S of Γ defines a word length metric d_S on Γ. The asymptotic structure of (Γ, d_S) was studied by Gromov [13] and Pansu [23]. For any sequence $r_i \to \infty$, Gromov-Hausdorff convergence holds:

$$(r_i^{-1} \Gamma, e, d_S) \xrightarrow{GH} (G, e, d).$$

The unique limit space (G, d) is a Carnot group, that is, a simply connected stratified nilpotent Lie group G with nilpotency step l and a distance d induced by a left-invariant subFinsler metric. Moreover, $\dim_{\mathcal{H}}(Ly) = l$ for any one-parameter subgroup L in $\zeta_{l-1}(G)$, the last non-trivial subgroup in the lower central series. This structure also applies to closed manifolds. For a closed Riemannian manifold (M, g) with a virtually nilpotent fundamental group Γ, although g cannot have nonnegative Ricci curvature when Γ has nilpotency step $\geqslant 2$, the blow-down sequence of the universal cover $(r_i^{-1} \widetilde{M}, \tilde{p}, \tilde{g})$ actually converges in the Gromov-Hausdorff topology to a limit space (G, e, d) as described above.

When it comes to open manifolds with $\mathrm{Ric} > 0$ and torsion-free nilpotent fundamental groups mentioned in Example 9, the asymptotic cone of the universal cover always has an isometric $\mathbb{R}$-orbit with Hausdorff dimension $\geqslant 2$; see [19, Section 3] for details.

There is indeed a relation between the nilpotency step of $\pi_1(M)$ and the Hausdorff dimension of $\mathbb{R}$-orbits in asymptotic cones [19]:

Theorem 33([19]) *Let (M, p) be an open n-manifold with $\mathrm{Ric} \geqslant 0$ and $E(M, p) \neq 1/2$. Let $\mathcal{N}$ be a torsion-free nilpotent subgroup of $\pi_1(M, p)$ with nilpotency step l and finite index. Then there exists an asymptotic cone (Y, y) of $\widetilde{M}$, the universal cover of M, and a closed $\mathbb{R}$-subgroup L of $\mathrm{Isom}(Y)$ such that*

$\dim_{\mathcal{H}}(Ly) \geqslant l$.

We say that $\widetilde{M}$ is conic at infinity, if any asymptotic cone (Y, y) of $\widetilde{M}$ is a metric cone with vertex y. For a metric cone (Y, y), any isometric $\mathbb{R}$-orbit at y must have Hausdorff dimension exactly 1. Thus Theorem 33 implies the virtual abelianness of π_1 in this case. When $\widetilde{M}$ has Euclidean volume growth, by a result of Cheeger-Colding, it is conic at infinity. We can also control the index of the abelian subgroup by the dimension and volume growth rate [18].

Theorem 34([18]) *Let (M, p) be an open n-manifold with $\mathrm{Ric} \geqslant 0$ and $E(M, p)$ $\neq 1/2$.*

(1) If its Riemannian universal cover is conic at infinity, then $\pi_1(M)$ is virtually abelian.

(2) If its Riemannian universal cover has Euclidean volume growth of constant at least L, then $\pi_1(M)$ has an abelian subgroup of index at most $C(n, L)$, a constant only depending on n and L.

We close this section with a question below, which may be related to Question 28 in the light of Theorem 33.

Question 35 *What is the smallest dimension n such that there is an open n-manifold with $\mathrm{Ric} \geqslant 0$ and a torsion-free nilpotent fundamental group?*

4 Topology of Ricci limit/RCD spaces

This chapter concerns the universal covers and semi-local simple connectedness of Ricci limit spaces or RCD spaces. In general, a Ricci limit space, even non-collapsing, may have infinite second or higher Betti number locally [12]. Recently, Hupp, Naber, and Wang have constructed a collapsing Ricci limit space such that any open set U has infinitely generated $H^2(U)$ [11]. Thus it is natural to ask whether a Ricci limit space is semi-locally simply connected, or equivalently, whether it has a simply connected universal cover.

4.1 Relative δ-covers

The universal cover is often defined as the simply connected cover. Here we do not assume it is simply connected, instead as the cover of all covers.

Definition 36([30, Page 82]) We say $\widetilde{X}$ is a universal cover of a path-connected space X if $\widetilde{X}$ is a cover of X such that for any other cover $\bar{X}$ of X, there is a commutative triangle formed by a covering map $f : \widetilde{X} \to \bar{X}$ and the two covering projections as below:

$$\widetilde{X} \quad \xrightarrow{\ f\ } \quad \bar{X}$$
$$\searrow \qquad \swarrow$$
$$X$$

Let $\mathcal{U}$ be any open covering of X. For any $x \in X$, by [30, Page 81], there is a covering space $\widetilde{X}_{\mathcal{U}}$ of X with covering group $\pi_1(X,\mathcal{U},p)$, where $\pi_1(X,\mathcal{U},x)$ is a normal subgroup of $\pi_1(X,p)$ generated by homotopy classes of closed paths having a representative of the form $\alpha^{-1} \circ \beta \circ \alpha$, where β is a closed path lying in some element of $\mathcal{U}$ and α is a path from x to $\beta(0)$.

Now we recall the notion of δ-covers introduced in [28] which plays an important role in studying the existence of the universal cover.

Definition 37 Given $\delta > 0$, the δ-cover, denoted $\widetilde{X}^\delta$, of a length space X is defined to be $\widetilde{X}_{\mathcal{U}_\delta}$, where $\mathcal{U}_\delta$ is the open covering of X consisting of all balls of radius δ.

Intuitively, a δ-cover is the result of unwrapping all but the loops generated by small loops in X. Clearly $\widetilde{X}^{\delta_1}$ covers $\widetilde{X}^{\delta_2}$ when $\delta_1 \leqslant \delta_2$.

Definition 38 (Relative δ-cover) Suppose X is a length space, $x \in X$ and $0 < r < R$. Let

$$\pi^\delta : \widetilde{B}_R(x)^\delta \to B_R(x)$$

be the δ-cover of the open ball $B_R(x)$. A connected component of

$$(\pi^\delta)^{-1}(B(x,r)),$$

where $B(x,r)$ is a closed ball, is called a relative δ-cover of $B(x,r)$ and is denoted $\widetilde{B}(x,r,R)^\delta$.

4.2 Universal cover of Ricci limit/RCD space exists

In [29, Lemma 2.4, Theorem 2.5], it is shown that if the relative δ-cover stabilizes, then the universal cover exists. This is the key tool for showing the existence of the universal cover in the works of Sormani and the second-named author.

Theorem 39 ([28, 29]) *Let (X, d) be a length space and assume that there is $x \in X$ with the following property: for all $r > 0$, there exists $R \geqslant r$, such that $\tilde{B}(x, r, R)^\delta$ stabilizes for all δ sufficiently small. Then (X, d) admits a universal cover $\tilde{X}$. More precisely $\tilde{X}$ is obtained as covering space $\tilde{X}_{\mathcal{U}}$ associated to a suitable open cover $\mathcal{U}$ of X satisfying the following property: for every $x \in X$ there exists $U_x \in \mathcal{U}$ such that U_x is lifted homomorphically by any covering space of (X, d).*

Theorem 40 ([28, 29]) *If X is the Gromov-Hausdorff limit of a sequence of complete Riemannian manifolds M_i^n with Ricci curvature $\geqslant K$, then X has a universal cover.*

These results on Ricci limit spaces were later generalized to RCD spaces by Mondino and the second-named author.

Theorem 41 ([14]) *Any $\mathsf{RCD}^*(K, N)$ space $(X, \mathsf{d}, \mathfrak{m})$ admits a universal cover $(\tilde{X}, \tilde{\mathsf{d}}, \tilde{\mathfrak{m}})$, which is itself $\mathsf{RCD}^*(K, N)$, where $K \in \mathbb{R}$, $N \in (1, +\infty)$.*

By Theorem 39, in order to prove that the universal exists, it is enough to show the relative covers stabilize. See [14, Theorem 4.5].

Theorem 42([14]) *Let $(X, \mathsf{d}, \mathfrak{m})$ be an $\mathsf{RCD}^*(K, N)$ space for some $K \in \mathbb{R}, N \in (1, \infty)$. For all $R > 0$ and $x \in X$, there exists $\delta_{x,R}$ depending on X, x, R such that*

$$\tilde{B}(x, \tfrac{R}{10}, R)^{\delta_{x,R}} = \tilde{B}(x, \tfrac{R}{10}, R)^\delta, \qquad \forall \delta < \delta_{x,R}.$$

Theorem 42 plays an important role in the work of Wang [32] showing that RCD spaces are semi-locally simply connected.

4.3 Ricci limit/RCD spaces are semi-locally simply connected

Recall that the universal cover $\tilde{X}$ is simply connected iff X is semi-locally simply connected, which means that there exists a neighbourhood such that every loop is contractible in X.

Definition 43(1-contractibility radius)

$$\rho(t, x) = \inf\{\infty, \rho \geqslant t | \text{ any loop in } B_t(x) \text{ is contractible in } B_\rho(x)\}.$$

X is semi-locally simply connected if for any $x \in X$, there is $T > 0$ such that $\rho(T, x) < \infty$.

In [20], for noncollapsing Ricci limit spaces, we show they are essentially locally simply connected.

Theorem 44 ([20]) *Any non-collapsing Ricci limit space is semi-locally simply connected. Therefore the universal cover is simply connected. In fact*

$$\lim_{t \to 0} \frac{\rho(t, x)}{t} = 1.$$

In the paper, we illustrate several ways of constructing homotopy. One way is to construct a homotopy by defining it on finer and finer skeletons of closed unit disk, see [20, Lemma 4.1].

The case for general Ricci limit spaces and RCD spaces are resolved by Wang [31, 32]. A key step by Wang shows that the stability of local δ-covers indeed implies semi-local simple connectedness.

Theorem 45 ([32]) *For a locally compact length metric space X, if any local relative δ-cover is stable, then X is semi-locally simply connected.*

Combining this with Theorem 42 it shows that the universal cover of RCD(k, N), where $N < \infty$, is simply connected.

To prove Theorem 45, the key is to use stability of the local relative δ-cover to show any loop in a small neighborhood of an RCD(K, N) space is homotopic to some loops in very small balls by a controlled homotopy image.

Lemma 46([32]) *For any $x \in (X, \mathsf{d}, \mathfrak{m})$, an RCD$(K, N)$, any $l < 1/2$, and small $\delta > 0$, there exist $\rho < l$ and $k \in \mathbb{N}$ such that any loop $\gamma \subset B_\rho(x)$ is homotopic to the union of some loops γ_i $(1 \leqslant i \leqslant k)$ in δ-balls and the homotopy image is in $B_{4l}(x)$.*

Applying the above lemma iteratively, one can construct the needed homotopy by using a similar method in [20, Lemma 4.1]. Namely, first shrink γ to smaller loops in δ_1-balls, then the second step is to shrink each new loop to even smaller loop in δ_2-balls, etc. Since the homotopy to shrink each loop is contained in a l_i-ball in the i-th step, this process converges to a homotopy map which contracts γ while the image is contained in a ball with radius $\sum\limits_{i=1}^{\infty} l_i \leqslant R$.

In terms of the 1-contractibility radius, the above argument indeed proves that $\rho(t, x) \to 0$ as $t \to 0$. It is unclear whether a stronger estimate $\rho(t, x)/t \to 1$

in the non-collapsing case holds in general.

Question 47 $\lim\limits_{t \to 0} \dfrac{\rho(t,x)}{t} = 1$ *?*

References

[1] Lionel Bérard-Bergery. *Quelques exemples de variétés riemanniennes complètes non compactes à courbure de Ricci positive.* C. R. Acad. Sci. Paris Sér. I Math., 302(4): 159–161, 1986.

[2] Igor Belegradek and Guofang Wei. *Metrics of positive Ricci curvature on vector bundles over nilmanifolds.* Geom. Funct. Anal., 12: 56–72, 2002.

[3] Igor Belegradek and Guofang Wei. *Metrics of positive Ricci curvature on bundles.* Inter. Math. Res. Not., 57: 3079–3096, 2004.

[4] Elia Brue, Aaron Naber, Daniele Semola. *Fundamental groups and the Milnor conjecture.* arXiv: 2303.15347.

[5] Jeff Cheeger and Tobias H. Colding. *Lower bounds on Ricci curvature and the almost rigidity of warped products.* Ann. Math., 144(1): 189–237, 1996.

[6] Jeff Cheeger and Tobias H. Colding. *On the structure of spaces with Ricci curvature bounded from below I, II, III.* J. Diff. Geom., 46(3): 406–480, 1997; 54(1): 13–35, 2000; 54(1): 37–74, 2000.

[7] Jeff Cheeger and Detlef Gromoll. *The splitting theorem for manifolds of nonnegative Ricci curvature.* J. Diff. Geom., 6: 119–128, 1971/72.

[8] Jeff Cheeger and Detlef Gromoll. *On the structure of complete manifolds of nonnegative curvature.* Ann. Math., 96: 413–443, 1972.

[9] Tobias H. Colding and Aaron Naber. *Sharp Hölder continuity of tangent cones for spaces with a lower Ricci curvature bound and applications.* Ann. Math., 176: 1173–1229, 2012.

[10] Xianzhe Dai, Shouhei Honda, Jiayin Pan, and Guofang Wei. *Singular Weyl' s law with Ricci curvature bounded.* Trans. Amer. Math. Soc. Ser. B, 10: 1212–1253, 2023.

[11] Erik Hupp, Aaron Naber, and Kai-Hsiang Wang. *Lower Ricci Curvature and Nonexistence of Manifold Structure.* arXiv: 2308.03909.

[12] Xavier Menguy. *Noncollapsing examples with positive Ricci curvature and infinite topological types.* Geom. Funct. Anal., 10: 600–627, 2000.

[13] Mikhael Gromov. *Groups of polynomial growth and expanding maps.* Inst. Hautes Études Sci. Publ. Math., 53: 53–78, 1981.

[14] Andrea Mondino and Guofang Wei. *On the universal cover and the fundamental group of an $RCD^*(K,N)$-space.* J. Reine Angew. Math., 753: 211–237, 2019.

[15] Philippe Nabonnand. *Sur les variétés riemanniennes complètes à courbure de Ricci positive.* C. R. Acad. Sci. Paris Sér. A-B, 291(10): A591–A593, 1980.

[16] Jiayin Pan. *On the escape rate of geodesic loops in an open manifold with nonnegative Ricci curvature.* Geom. Topol., 25(2): 1059–1085, 2021.

[17] Jiayin Pan. *Nonnegative Ricci curvature and escape rate gap.* J. Reine Angew. Math., 782: 175–196, 2022.

[18] Jiayin Pan. *Nonnegative Ricci curvature, metric cones, and virtual abelianness.* To appear in Geom. Topol., arXiv: 2201.07852.

[19] Jiayin Pan. *Nonnegative Ricci curvature, nilpotency, and Hausdorff dimension.* arXiv: 2309.01147

[20] Jiayin Pan and Guofang Wei. *Semi-local simple connectedness of noncollapsing Ricci limit spaces.* J. Eur. Math. Soc., 24(12): 4027–4062, 2022.

[21] Jiayin Pan and Guofang Wei. *Examples of Ricci limit spaces with non-integer Hausdorff dimension.* Geom. Funct. Anal., 32(3): 676–685, 2022.

[22] Jiayin Pan and Guofang Wei. *Examples of open manifolds with positive Ricci curvature and non-proper Busemann functions.* arXiv: 2203.15211.

[23] Pierre Pansu. *Croissance des boules et des géodésiques fermées dans les nilvariétés.* Ergod. Th. & Dynam. Sys., 3(3): 415–445, 1983.

[24] Vladimir A. Sharafutdinov. *On convex sets in a manifold of nonnegative curvature.* Math. Notices, 26(1): 129–136, 1979.

[25] Zhongmin Shen. *Complete manifolds with nonnegative Ricci curvature and large volume growth.* Invent. Math., 125(3): 393–404, 1996.

[26] Christina Sormani. *Busemann functions on manifolds with lower bounds on Ricci curvature and minimal volume growth.* J. Diff. Geom., 48(3): 557–585, 1998.

[27] Christina Sormani. *Ricci curvature, small linear diameter growth, and finite generation of fundamental groups.* J. Diff. Geom., 54(3): 547–559, 2000.

[28] Christina Sormani and Guofang Wei. *Hausdorff convergence and universal covers.* Trans. Amer. Math. Soc., 353: 3585–3602, 2001.

[29] Christina Sormani and Guofang Wei. *Universal covers for Hausdorff limits of noncompact spaces.* Trans. Amer. Math. Soc., 356(3): 1233–1270, 2004.

[30] Edwin H. Spanier. *Algebraic Topology.* McGraw-Hill, Inc., 1966.

[31] Jikang Wang. *Ricci limit spaces are semi-locally simply connected.* arXiv: 2104.02460.

[32] Jikang Wang. *$RCD^*(k, N)$ spaces are semi-locally simply connected.* To appear in J. Reine Angew. Math., arXiv: 2211.07087.

[33] Guofang Wei. *Examples of complete manifolds of positive Ricci curvature with nilpotent isometry groups.* Bull. Amer. Math. Soc. (N.S.), 19(1): 311–313, 1988.

Harmonic Metrics and Higgs Bundles

Di Wu

School of Mathematics and Statistics, Nanjing University of Science and Technology, Nanjing, China

E-mail: wudi@njust.edu.cn

Abstract   Harmonic metrics emerge as canonical metrics on vector bundles, this survey paper gives an exposition of our recent results on the existence criteria for harmonic metrics as well as related topics on Higgs bundles over noncompact Kähler manifolds and compact Sasakian manifolds.

1   Introduction

Throughout this paper, unless indicated explicitly otherwise, vector bundles could be real or complex and whose metrics are Riemannian or Hermitian respectively. Let E be a vector bundle over a compact Riemannian manifold (M, g), equipped with a connection ∇ on E. Naturally, there arises a basic problem: *What is the best canonical metric on (E, ∇)?*

Motivated by calculating the Euler characteristic number of E via the Gauss-Bonnet-Chern formula and its ramifications, one may ask: *How far is ∇ from being metric compatible?* For this purpose, given a metric K on E, we write

$$\nabla = \nabla_K + \psi_K, \tag{1.1}$$

where ∇_K is a connection preserving K and $\psi_K \in A^1(\text{End}(E))$. Consider the

The author is supported by the project funded by China Postdoctoral Science Foundation 2023M731699 and the Jiangsu Funding Program for Excellent Postdoctoral Talent 2022ZB282.

following functional on the space of metrics:

$$\mathcal{E}_\nabla(K) = \frac{1}{2} \int_M |\psi_K|^2 \, \mathrm{dvol}_g, \tag{1.2}$$

and the point is to minimize $\mathcal{E}_\nabla(\bullet)$. The canonical metric we are interested in is the critical point of $\mathcal{E}_\nabla(\bullet)$, which satisfies the tensor-valued equation

$$\nabla_H^* \psi_H = 0, \tag{1.3}$$

and we call such H a harmonic metric. There is a reason for the name of harmonic metric as follows. If ∇ is a flat connection, any metric H on E corresponds to an equivariant map $f_H : \tilde{M} \to GL(r)/U(r)$, where $\tilde{M}$ is the universal covering space of M and $r = \mathrm{rank}(E)$. It is known that H being a harmonic metric iff f_H being a harmonic map. Hence, one may study harmonic metrics from the viewpoint of harmonic maps and the flatness assumption on ∇ plays a similar role as that of the nonpositive curvature assumption on $GL(r)/U(r)$. In general, lacking of flatness may block off methods used in equivariant harmonic maps.

Another important reason for considering harmonic metrics as canonical metrics on vector bundles lies in the scenario of nonabelian Hodge theory in Kähler geometry. Recall that for a compact Kähler manifold X, the Hodge decomposition indicates

$$H_{dR}^1(X, \mathbb{C}) \cong H_{Dol}^{0,1}(X, \mathbb{C}) \oplus H_{Dol}^{1,0}(X, \mathbb{C}). \tag{1.4}$$

This isomorphism may be also explained as: a flat line bundle can be identified to a holomorphic line bundle with vanishing Chern classes and a holomorphic form of type (1.0). In higher rank case, via the theory of harmonic metrics, the Corlette-Simpson correspondence [Co1, Simi1, Simi3](also referred as the nonabelian Hodge correspondence, or the Donaldson-Hitchin correspondence [Do3, Hi] when $\dim_{\mathbb{C}} X = 1$) states that there is an equivalence between the category of semi-simple flat complex vector bundles(analogous to the abelian de Rham cohomology $H_{dR}^1(X, \mathbb{C})$) and the category of poly-stable Higgs bundles with vanishing Chern classes(analogous to the abelian Dolbeault cohomology $H_{Dol}^{0,1}(X, \mathbb{C}) \oplus H_{Dol}^{1,0}(X, \mathbb{C})$).

The rest of this survey paper is organized as follows. In Section 2, we survey existence criteria for harmonic metrics. In Section 3, we apply harmonic metrics to construct Higgs bundles on noncompact Kähler manifolds, intending to investigate Corlette-Simpson correspondence of noncompact type. In Section 4, we discuss Sasakian Corlette-Simpson correspondence in which particular emphasis will be placed on an important characterization of harmonic metrics in Sasakian geometry.

2   Existence criteria for harmonic metrics

The first step to study harmonic metrics is to attack the existence problem. A unifying principle in geometric analysis predicts that the existence of canonical metrics should be closely related to appropriate stability conditions on underlying geometric objects, one may refer to the Yau-Tian-Donaldson problem for extremal metrics in Kähler geometry and the Donaldson-Uhlenbeck-Yau characterization(also referred as the Hitchin-Kobayashi correspondence or the Kobayashi-Hitchin correspondence) for Hermitian-Yang-Mills metrics on holomorphic vector bundles.

Detecting harmonic metrics on vector bundles is a nonlinear system generalization of solving Laplace equations and obstruction would appear. To illustrate related results, let us recall

Definition 1   A vector bundle (E, ∇) is said to be semi-simple(also called completely reducible, reductive, or poly-stable) if it splits as a direct sum of simple sub-bundles, where simple(also called irreducible or stable) means that there exists no non-trivial ∇-invariant sub-bundle.

Based on the Codazzi-Gauss type equation for vector bundles, it is easy to conclude the semi-simpleness if there exists a harmonic metric. In the late 1980s, borrowing ideas from harmonic maps, Donaldson [Do3] proved that any semi-simple flat principal $GL(2, \mathbb{C})$-bundle over a compact Riemannian surface must admit harmonic metrics. Deforming a semi-simple flat connection via an evolution equation and utilizing Uhlenbeck compactness [Uh], it is Corlette [Co1] who proved the existence of harmonic metrics on a semi-simple flat vector bundle over a compact Riemannian manifold. Precisely, we have

Theorem 2(Corlette-Donaldson theorem)   *Let (E, ∇) be a vector bundle over a compact Riemannian manifold with a flat connection ∇, then it admits harmonic*

metrics if and only if it is semi-simple.

Many works are engaged in generalizations on Corlette-Donaldson's tremendous criterion since then and a main reason is to further establish Corlette-Simpson type correspondences in different contexts. For example, if (E, ∇) is non-Hermitian Yang-Mills(that is, the curvature F_∇ satisfies $F_\nabla \in A^{1,1}(\text{End}(E))$ and $\Lambda F_\nabla = c \, \text{id}_E$ for a constant c, where Λ is the operator obtained as the adjoint of the multiplication of the Kähler form) instead of flatness and the base space is a compact Kähler manifold, Kaledin-Verbitsky [KV] conjectured that the existence of harmonic metrics is equivalent to a kind of stability condition, see [PSZ]. The original motivation to consider non-Hermitian Yang-Mills bundles lies on an attempt to investigate Corlette-Simpson correspondence on vector bundles with arbitrary Chern classes but which seems still not to be fully understood. On the other hand, to study Corlette-Simpson correspondence on non-Kähler manifolds, it is natural to consider the following modified harmonic metric equation on flat complex vector bundles:

$$\Lambda G_H = 0, \; G_H = (\nabla_H^{0,1} + \psi_H^{1,0})^2, \tag{2.1}$$

which is equivalent to the harmonic metric equation (1.3) in Kähler case. Readers can consult [Lü, PZZ] for existence results on (2.1) and non-Kähler Corlette-Simpson correspondence.

Recently, we extended the Corlette-Donaldson theorem to arbitrary vector bundles without any additional hypothesis. In fact, we proved

Theorem 3([WZ2]) *Let (E, ∇) be a vector bundle over a compact Riemannian manifold with an arbitrary connection ∇, then it admits harmonic metrics if and only if it is semi-simple.*

Theorem 3 is an analogue of the theory of Hermitian-Yang-Mills metrics in Kähler geometry [Do1, Do2, NS, UY]. In Riemannian category, we perform various calculations of the harmonic metric equation on vector bundles equipped with arbitrary connections. Beyond that, a priori estimates along the continuity method are based on a regularity lemma on weakly ∇-invariant sub-bundles which serves as a smooth vector bundle counterpart of Uhlenbeck-Yau's deep result on weakly holomorphic sub-bundles [UY]. Since it deals with the real operator ∇ rather than $\overline{\partial}$, here one can avoid the intricate proof of Uhlenbeck-

Yau. Note Loftin [Lo] had also considered the similar issue in his study of Donaldson-Uhlenbeck-Yau type theorem for flat vector bundles over special affine manifolds.

Below we consider the existence problem in noncompact setting.

Inspired by [Simi1, Mo2], we list the following assumptions.

Assumption 4   *It admits a function $\phi_M : M \to \mathbb{R}_{\geqslant 0}$ with $\phi_M \in L^1$, such that if f is a nonnegative bounded function satisfying $\Delta f \geqslant -B\phi_M$ for a positive constant B in the distribution sense, we have*

$$\sup_M f \leqslant C(B)(1 + \int_M f\phi_M \, \mathrm{dvol}_g). \tag{2.2}$$

Furthermore, if f satisfies $\Delta f \geqslant 0$, we have $\Delta f = 0$.

Assumption 5   *It admits an exhaustion function $\rho : M \to \mathbb{R}_{\geqslant 0}$ with $|\Delta \rho| \in L^1$.*

As Simpson did in the case of Hermitian-Yang-Mills metrics [Simi1], we define

$$\deg_g(E, \nabla, K) = -\int_M \operatorname{tr} \nabla_K^* \psi_K \, \mathrm{dvol}_g . \tag{2.3}$$

Definition 6   If it holds for any non-trivial ∇-invariant sub-bundle $S \subseteq E$ that

$$\frac{\deg_g(S, \nabla|_S, K|_S)}{\operatorname{rank}(S)} < \frac{\deg_g(E, \nabla, K)}{\operatorname{rank}(E)}, \tag{2.4}$$

then (E, ∇, K) is said to be stable.

Remark 7   *If M is compact, (E, ∇, K) being stable(poly-stable) iff (E, ∇) being simple(semi-simple), while the correct substitution should be the above analytic stability(it depends on the choice of metric K) in noncompact setting since we shall prove it is equivalent to the existence of special harmonic metrics. If (M, g) is a Kähler manifold and ∇ is flat, our definition coincides with that in [Simi2].*

The following can be interpreted as a noncompact Corlette-Donaldson theorem.

Theorem 8 ([WZ1])   *Let (E, ∇, K) be a vector bundle over a noncompact Riemannian manifold (M, g) satisfying Assumptions 4, 5 with $\phi_M = 1$, with a flat connection ∇ such that there is no non-trivial parallel section of $\operatorname{End}(E)$*

and a metric K satisfying $|\nabla_K^* \psi_K| \in L^\infty$, $|\psi_K| \in L^2$. Then the following two statements are equivalent:

1. (E, ∇, K) is stable.
2. (E, ∇, K) admits a unique harmonic metric H with $\det K = \det H$, $|h| \in L^\infty$ and $|\nabla h| \in L^2$, where h is given by $H(\cdot, \cdot) = K(h(\cdot), (\cdot))$.

Concerning our result, we remark that the direction from stability to existence should be well-understood for stable flat complex vector bundles over noncompact Kähler manifolds after [Simi1, Simi2]. In fact, Simpson had pointed out in [Simi2] that the heat flow argument used in detecting Hermitian-Yang-Mills metrics [Simi1] can be adapted to find harmonic metrics(also see [Mo1] for its exposition). However, their method of deriving a prior estimates is not directly applicable in Riemannian setting due to the absence of certain functionals. For the opposite direction, we stress that the stability proved is with respect to the background data K. This and the uniqueness part both require comparisons of different stabilities. It may be mentioned that this property is facilitated by the feature of the stability of vector bundles, while analogous property is generally unknown in the context of Hermitian-Yang-Mills metrics.

For more results related to the existence problem on harmonic metrics in noncompact case, readers are referred to [CJY, Co2, JZ1, JZ2, Li, Mo1].

3 Constructing Higgs bundles on noncompact manifolds

Definition 9 A Higgs bundle consists of a holomorphic vector bundle $(E, \bar{\partial}_E)$ and a holomorphic section $\theta \in A^{1,0}(\mathrm{End}(E))$ such that $\theta \wedge \theta = 0$. Then θ is called a Higgs field and $(\bar{\partial}_E, \theta)$ is a Higgs structure on the underlying vector bundle E.

Definition 10 For a Higgs bundle $(E, \bar{\partial}_E, \theta)$ and a Hermitian metric K, we define

$$\nabla_{E,\theta,K} = \partial_{\theta,K} + \bar{\partial}_{E,\theta}, \quad \partial_{\theta,K} = \partial_K + \theta^{*K}, \quad \bar{\partial}_{E,\theta} = \bar{\partial}_E + \theta, \qquad (3.1)$$

where $\nabla_{E,K} = \partial_K + \bar{\partial}_E$ is the Chern connection. We say a flat complex vector bundle (E, ∇) comes form a Higgs bundle if $\nabla = \nabla_{E,\theta,H}$ for some $(\bar{\partial}_E, \theta)$ and H.

Constructing Higgs structures via harmonic metrics is a key step to establish Corlette-Simpson correspondence. If X is compact and H being harmonic metric, this is equivalent to prove the pluriharmonicity of H, the issue does not appear and it is essentially due to Siu-Sampson [Sam, Siu] and Corlette [Co1], also see [Simi3]. If X is noncompact, see [Simi1, JZ2] for the case that the complement of a divisor in a compact Kähler manifold. We also present a different approach to construct Higgs bundles on noncompact manifolds. The result is as follows.

Theorem 11 ([WZ1])　*Let (X, ω) be a Kähler manifold satisfying Assumptions 4, 5 with $\phi_M = 1$ and (E, ∇, K) be a stable flat complex vector bundle over X such that $|\nabla_K^* \psi_K| \in L^\infty$ and $|\psi_K| \in L^2$. Assume either $\dim_{\mathbb{C}} X = 1$ or (X, ω) being complete with bounded Ricci curvature below, then (E, ∇) comes from a Higgs bundle.*

By Theorem 8, we know that there exists a harmonic metric H and then the proof of Theorem 11 reduces to conclude $G_H = 0$, this can be proved by showing $|\nabla_H \psi_H| \in L^2$ (where ∇_H also denotes the connection on $T^* X \otimes \mathrm{End}(E)$ which acts on $T^* X$ by the Riemannian connection and acts on $\mathrm{End}(E)$ by ∇_H.) and applying Yau's lemma in [Ya].

Furthermore, the following theorem tells us when the Higgs field vanishes. Especially, it yields the existence of metric compatible flat connections on E and it is a harmonic metric version of the Liouville theorem in [SY].

Theorem 12 ([WZ1])　*Let (X, ω) be a complete noncompact Kähler manifold with nonnegative Ricci curvature and satisfy Assumption 4. Assume (E, ∇, K) is a stable flat complex vector bundle over X such that*

$$\left| \nabla_K^* \psi_K - \frac{\mathrm{tr}\, \nabla_K^* \psi_K}{\mathrm{rank}(E)} \, \mathrm{id}_E \right| \leqslant C \phi_M, \quad |\psi_K| \in L^2, \tag{3.2}$$

for a constant C, then ∇ is metric compatible.

As a generalization of the Corlette-Simpson correspondence to noncompact case, it should be interesting but challenging to establish a bijective correspondence between suitable spaces of flat bundles and Higgs bundles on some noncompact manifolds. One may see [WZ1] for a partial study on this problem.

4 Corlette-Simpson correspondence in Sasakian geometry

Sasakian manifolds were first introduced in [Sas], which is a subject lying on the intersection of contact, CR, Kähler and Riemannian geometry. A basic model of a compact Sasakian manifold is the odd dimensional unit sphere. We start with some basic facts on Sasakian manifolds from the viewpoint of pseudo-Hermitian geometry, more details can be found in [BG,DT]. Let M be a $2n+1$-dimensional smooth manifold, a CR structure on M is an integrable rank n complex sub-bundle

$$T_{1,0}M \subseteq T^{\mathbb{C}}M = TM \otimes \mathbb{C} \qquad (4.1)$$

satisfying $T_{1,0}M \cap T_{0,1}M = \{0\}$ for $T_{0,1}M = \overline{T_{1,0}M}$. We call $(M, T_{1,0}M)$ a CR manifold and its maximal complex or Levi, distribution is the rank $2n$ real sub-bundle

$$HM = Re\{T_{1,0}M \oplus T_{0,1}M\} \subseteq TM. \qquad (4.2)$$

It carries a complex structure $J : HM \to HM$ given by

$$J(X + \overline{X}) = \sqrt{-1}(X - \overline{X}), \ X \in T_{1,0}M. \qquad (4.3)$$

Assume M to be orientable, there is an orientable real line bundle E with fiber

$$E_x = \{\omega \in T_x^*M | HM_x \subseteq \ker \omega\} \subseteq T_x^*M, \qquad (4.4)$$

A globally defined nowhere vanishing section η will be called a pseudo-Hermitian structure and the associated Levi form L_η is defined by

$$L_\eta(X, \overline{Y}) = -\sqrt{-1}d\eta(X, \overline{Y}), \ X, Y \in T_{1,0}M. \qquad (4.5)$$

An orientable CR manifold $(M, T_{1,0}M)$ is nondegenerate if L_η is nondegenerate for a pseudo-Hermitian structure η and then $(M, T_{1,0}M, \eta)$ is called a pseudo-Hermitian manifold. Moreover, $(M, T_{1,0}M, \eta)$ is said to be strictly pseudoconvex CR if L_η is positive definite.

For a pseudo-Hermitian manifold $(M, T_{1,0}M, \eta)$, $d\eta$ is nondegenerate on HM and thus there is a unique nonvanishing vector field ξ(referred to as the Reeb vector field or the characteristic direction) such that

$$\eta(\xi) = 1, \ d\eta(\xi, \bullet) = 0, \qquad (4.6)$$

and ξ is transverse to the Levi distribution(that is $TM = HM \oplus \mathbb{R}\xi$). Define a bilinear form

$$G_\eta(X, Y) = d\eta(X, JY), \ X, Y \in HM, \tag{4.7}$$

the integrability of $T_{1,0}M$ implies that G_η is J-invariant and thus symmetric. One may extend G_η to a semi-Riemannian(Riemannian if strictly pseudoconvex CR) metric via

$$g_\eta(X, Y) = G_\eta(X, Y), \ g_\eta(X, \xi) = 0, \ g_\eta(\xi, \xi) = 1, X, Y \in HM, \tag{4.8}$$

and we call it the Webster metric.

Proposition 13 ([Ta, We], also see [DT]) *For a pseudo-Hermitian manifold* $(M, T_{1,0}M, \eta)$, *we extend* J *to an endomorphism of* TM *by requiring* $J\xi = 0$. *There exists a unique affine connection* ∇^{TM} *(called the Tanaka-Webster connection) on* TM *such that*

    1. HM is parallel with respect to ∇^{TM},

    2. $\nabla^{TM} g_\eta = 0$, $\nabla^{TM} J = 0$, $\nabla^{TM}\xi = 0$ and $\nabla^{TM} d\eta = 0$,

    3. The torsion $T_{\nabla^{TM}}$ is pure:

    (a) $\tau_{\nabla^{TM}} \circ J + J \circ \tau_{\nabla^{TM}} = 0$,

    (b) $T_{\nabla^{TM}}(X, Y) = 0$, $T_{\nabla^{TM}}(X, \overline{Y}) = d\eta(X, \overline{Y})\xi$,

    where $\tau_{\nabla^{TM}} = T_{\nabla^{TM}}(\xi, \bullet)$(called the pseudo-Hermitian torsion) and $X, Y \in T_{1,0}M$.

Definition 14 A strictly pseudoconvex CR manifold $(M, T_{1,0}M, \eta)$ with vanishing pseudo-Hermitian torsion will be called a Sasakian manifold.

The Corlette-Simpson correspondence in Sasakian geometry reads as follows.

Theorem 15 ([BK1]) *For a compact Sasakian manifold* $(M, T_{1,0}M, \eta)$, *there exists an equivalence between the following two categories:*

- *Semi-simple flat complex vector bundles over M.*
- *Poly-stable basic Higgs bundles over M with vanishing basic Chern classes.*

In [BM], Theorem 15 was previously investigated for the special case of quasi-regular Sasakian manifolds. We shall discuss key ingredients in proving

Sasakian Corlette-Simpson correspondence. The direction from basic Higgs bundles to flat bundles is mainly inspired by a foliated Hitchin-Kobayashi correspondence [BH], where a notion of stability of basic holomorphic bundles was addressed in order to get around the problem of absence of basic Hermitian metrics on a determinant line bundle associated with a sub-sheaf. A central point in the opposite direction is based on the Reeb invariance of harmonic metrics in Sasakian geometry as follows.

Theorem 16 ([BK1]) *Let (E, ∇) be a flat vector bundle over a compact Sasakian manifold $(M, T_{1,0}M, \eta)$ and H be a harmonic metric, then H is Reeb invariant.*

In [BK1], Theorem 16 was observed by modifying the spinorial trick developed by Petit [Pe] in his study of harmonic maps on strictly pseudoconvex CR manifolds. It should be compared to [Pe, Theorem 4.1] that any harmonic map from a compact Sasakian manifold to a Riemannian manifold with nonpositive curvature must be foliated which means that the tangent map vanishes in the characteristic direction. For the spinorial proof, the authors in [BK1] also mentioned that this is facilitated by a special feature of Sasakian geometry which would not hold for a general transversally Kähler geometry.

In [WZ2], we presented a new proof of Theorem 16 under weaker condition.

Theorem 17 ([WZ2]) *Let (E, ∇) be a flat vector bundle over a compact Sasakian manifold $(M, T_{1,0}M, \eta)$ and H be a metric on E, then we have*

$$\nabla_H \nabla_H^* \psi_H = 0 \Leftrightarrow H \text{ is a Reeb invariant harmonic metric.}$$

In comparison with that in [BK1], we remark that our argument only involves the definition of Sasakian manifold in the sense of Definition 14, but does not involve two special properties of Sasakian manifolds: the curvature theory of Tanaka-Webster connections and the commutative formula [Pe, Corollary 2.1] concerning certain Dirac operators. One motivation to find such a simpler proof is to further study the foliated Corlette-Simpson correspondence.

We end this paper with an overview of the proof of Theorem 17. Let $(x^1, ..., x^{2n+1})$ be a local coordinate near the considered point and assume that $\{e_1 = \dfrac{\partial}{\partial x^1} = \xi, ..., e_{2n+1} = \dfrac{\partial}{\partial x^{2n+1}}\}$ locally.

Step 1: We observe the symmetry

$$\nabla_{H,e_i}(\psi_H(\xi)) = \nabla_{H,\xi}(\psi_H(e_i)), \tag{4.9}$$

where we used $\nabla^{TM}\xi = 0$, $\tau_{\nabla^{TM}} = 0$ and the flatness of ∇.

Step 2: Using (4.9), $\nabla^{TM}\xi = 0$, $\tau_{\nabla^{TM}} = 0$ and the fact $\mathrm{tr}_{g_\eta}(\nabla^{g_\eta} - \nabla^{TM}) = 0$, we prove

$$\mathrm{tr}_{g_\eta} \nabla_H^2(\psi_H(\xi)) = -\nabla_{H,\xi}\nabla_H^*\psi_H - \frac{1}{2}\mathrm{tr}_{g_\eta}[[\psi_H, \psi_H(\xi)], \psi_H]. \tag{4.10}$$

Furthermore, (4.10) and the assumption $\nabla_H\nabla_H^*\psi_H = 0$ imply

$$\Delta_{g_\eta}|\psi_H(\xi)|_H^2 = |[\psi_H(\xi), \psi_H]|_H^2 + 2|\nabla_H(\psi_H(\xi))|_H^2. \tag{4.11}$$

Hence the maximum principle gives

$$|\psi_H(\xi)|_H = constant, \ [\psi_H(\xi), \psi_H] = 0, \ \nabla_H(\psi_H(\xi)) = 0. \tag{4.12}$$

Step 3: Define the Dirac operator $\mathcal{D} = g_\eta^{ij}\left(e^i \wedge \nabla_{H,e_j} - \iota(e_i)\nabla_{H,e_j}\right)$ on $\Lambda^*T^*M \otimes \mathrm{End}(E)$, where for notational simplicity, ∇_H also denotes the connection on $\Lambda^*T^*M \otimes \mathrm{End}(E)$ which acts on Λ^*T^*M by ∇^{TM} and acts on $\mathrm{End}(E)$ by ∇_H. By $\mathrm{tr}_{g_\eta}(\nabla^{g_\eta} - \nabla^{TM}) = 0$ and the flatness of ∇, we prove

$$\mathcal{D}\psi_H = d\eta \otimes \psi_H(\xi) + \nabla_H^*\psi_H, \tag{4.13}$$

from which and (4.12), $\nabla^{TM}d\eta = 0$, $\nabla_H\nabla_H^*\psi_H = 0$, we find $\mathcal{D}^2\psi_H = 0$. Thus $\mathcal{D}\psi_H = 0$ since $\mathcal{D}$ is formal self-adjoint(see [Pe, Proposition 2.1]). In particular, by

$$\nabla_H^*\psi_H = 0, \ \psi_H(\xi) = 0. \tag{4.14}$$

Step 4: We may choose a local flat frame of E and it follows $\psi_H = -\frac{1}{2}h^{-1}dh$, where h is the local Hermitian matrix function corresponding to H. Finally, we conclude H is Reeb invariant by (4.14), as required.

Acknowledgments

Part of this paper is based on a talk on 2023 China Geometric Analysis Annual Conference at Xinjiang Normal University, I thank the organizers. This paper was also written when the author made a preparation for a talk on Geometry Seminar at Osaka University, I am grateful to Hisashi Kasuya for the academic visit invitation and the financial support on the travel, to Osaka University for the hospitality.

References

[BH]   David Baraglia and Pedram Hekmati. *A foliated Hitchin-Kobayashi corres-pondence*. Advances in Mathematics, 408(2022), part B, Paper No. 108661, 47 pp.

[BK1]   Indranil Biswas and Hisashi Kasuya. *Higgs bundles and flat connections over compact Sasakian manifolds*. Communications in Mathematical Physics, 385(2021), no. 1, 267-290.

[BM]   Indranil Biswas and Mahan Mj. *Higgs bundles on Sasakian manifolds*. International Mathematics Research Notices, 2018(2018), no. 11, 3490-3506.

[BG]   Charles P. Boyer and Krzysztof Galicki. *Sasakian Geometry*. Oxford Mathematical Monographs. Oxford: Oxford University Press, 2008.

[CJY]   Tristan C. Collins, Adam J. Jacob and Shing-Tung Yau. *Poisson metrics on flat vector bundles over non-compact curves*. Communications in Analysis Geometry, 27(2019), no. 3, 529-597.

[Co1]   Kevin Corlette. *Flat G-bundles with canonical metrics*. Journal of Differential Geometry, 28(1988), no. 3, 361-382.

[Co2]   Kevin Corlette. *Archimedean superrigidity and hyperbolic geometry*. Annals of Mathematics, 135(1992), no. 1, 165-182.

[Do1]   Simon K. Donaldson. *Anti self-dual Yang-Mills connections over complex algebraic surfaces and stable vector bundles*. Proceedings of the London Mathematical Society, 50(1985), no. 1, 1-26.

[Do2]   Simon K. Donaldson. *Infinite determinants, stable bundles and curvature*. Duke Mathematical Journal, 54(1987), no. 1, 231-247.

[Do3]   Simon K. Donaldson. *Twisted harmonic maps and the self-duality equations*. Proceedings of the London Mathematical Society, 55(1987), no. 1, 127-131.

[DT]   Sorin Dragomir and Giuseppe Tomassini. *Differential Geometry and Analysis on CR Manifolds*. Progress in Mathematics, 246. Boston: Birkhäuser Boston, Inc.

[Hi]   Nigel J. Hitchin. *The self-duality equations on a Riemann surface*. Proceedings of the London Mathematical Society, 55(1987), no. 1, 59-126.

[JZ1]   Jürgen Jost and Kang Zuo. *Harmonic maps and Sl(r, $\mathbb{C}$)-representations of fundamental groups of quasiprojective manifolds*. Journal of Algebraic Geometry, 5(1996), no. 1, 77-106.

[JZ2]   Jürgen Jost and Kang Zuo. *Harmonic maps of infinite energy and rigidity results for representations of fundamental groups of quasiprojective varieties.* Journal of Differential Geometry, 47(1997), no. 3, 469-503.

[KV]   Dmitry. Kaledin and Misha Verbitsky. *Non-Hermitian Yang-Mills connections.* Selecta Mathematica-New Series, 4(1998), no. 2, 279-320.

[Li]   Jiayu Li. *Hitchin's self-duality equations on complete Riemannian manifolds.* Mathematische Annalen, 306(1996), no. 3, 419-428.

[Lo]   John C. Loftin. *Affine Hermitian-Einstein metrics.* Asian Journal of Mathematics, 13(2009), no. 1, 101-130.

[Lü]   Martin Lübke. *Einstein metrics and stability for flat connections on compact Hermitian manifolds, and a correspondence with Higgs operators in the surface case.* Documenta Mathematica, 4(1999), 487-512.

[Mo1]   Takuro Mochizuki. *Kobayashi-Hitchin correspondence for tame harmonic bundles. II.* Geometry & Topology, 13(2009), no. 1, 359-455.

[Mo2]   T. Mochizuki. *Kobayashi-Hitchin correspondence for analytically stable bundles.* Transcations of the American Mathematical Society, 373(2020), no. 1, 551-596.

[NS]   Mudumbai S. Narasimhan and Conjeeveram S. Seshadri. *Stable and unitary vector bundles on a compact Riemann surface.* Annals of Mathematics, 82(1965), 540-567.

[PSZ]   Changpeng Pan, Zhenghan Shen and Xi Zhang. *On the existence of harmonic metrics on non-Hermitian Yang-Mills bundles.* arXiv 2301.01428, 2023.

[PZZ]   Changpeng Pan, Chuanjing Zhang and Xi Zhang. *Semi-stable Higgs bundles and flat bundles over over non-Kähler manifolds.* arXiv: 1911.03593v1, 2019.

[Pe]   Robert Petit. *Harmonic maps and strictly pseudoconvex CR manifolds.* Communications in Analysis Geometry, 10(2002), no. 3, 575-610.

[Sam]   Joseph H. Sampson. *Applications of harmonic maps to Kähler geometry//Yum-Tong Sim. Complex Differential Geometry and Nonlinear Differential Equations,* 125-134. Providence: Amer. Math. Soc., 1986.

[Sas]   Shigeo Sasaki. *On differentiable manifolds with certain structures which are closely related to almost contact structure. I.* Tohoku Mathematical Journal, 12(1960), 459-476.

[SY]   Richard M. Schoen and Shing-Tung Yau. *Harmonic maps and the topology of stable hypersurfaces and manifolds with non-negative Ricci curvature.* Commentarii Mathematici Helvetici, 51(1976), no. 3, 333-341.

[Simi1]   Carlos T. Simpson. *Constructing variations of Hodge structure using Yang-Mills theory and applications to uniformization.* Journal of the Amerian Mathematical Society, 1(1988), no. 4, 867-918.

[Simi2]   Carlos T. Simpson. *Harmonic bundles on noncompact curves.* Journal of the Amerian Mathematical Society, 3(1990), no. 3, 713-770.

[Simi3] Carlos T. Simpson. *Higgs bundles and local systems*. Publications Mathématiques de l'IHÉS, 75(1992), 5-95.

[Siu] Yum-Tong Siu. *The complex-analyticity of harmonic maps and strong rigidity of complex Kähler manifolds*. Annals of Mathematics, 112(1980), no. 1, 73-111.

[Ta] Noboru Tanaka. *A differential geometric study on strongly pseudo-convex manifolds*. Lectures in Mathematics, Department of Mathematics, Kyoto University, No. 9. Tokyo: Kinokuniya Book Store Co., Ltd., 1975.

[Uh] Karen K. Uhlenbeck. *Connections with L^p bounds on curvature*. Communications in Mathematical Physics, 83(1982), no. 1, 31-42.

[UY] Karen K. Uhlenbeck and Shing-Tung Yau. *On the existence of Hermitian-Yang-Mills connections in stable vector bundles*. Communications on Pure and Applied Mathematics, 39(1986), no. S, Suppl., S257-S293.

[We] Sidney M. Webster. *Pseudo-Hermitian structures on a real hypersurface*. Journal of Differential Geometry, 13(1978), no. 1, 25-41.

[WZ1] Di Wu and Xi Zhang. *Poisson metrics and Higgs bundles over noncompact Kähler manifolds*. Calculus of Variations and Partial Differential Equations, 62(2023), no. 1, Paper No. 20, 29 pp.

[WZ2] Di Wu and Xi Zhang. *Harmonic metrics and semi-simpleness*. arXiv: 2304.10800, 2023.

[Ya] Shing-Tung Yau. *Some function-theoretic properties of complete Riemannian manifold and their applications to geometry*. Indiana University Mathematics Journal, 25(1976), no. 7, 659-670.

Pluriclosed Flow on Hermitian-Symplectic Manifolds

YANAN YE

Beijing International Center for Mathematical Research, Peking University, Beijing, China

E-mail: yeyanan@stu.pku.edu.cn

Abstract This is a note on the pluriclosed flow with Hermitian-symplectic initial data. More generally, we consider this flow beginning with a pluriclosed metric with torsion potential. Under this assumption, we can find a monotonic quantity along the flow. As an application, we show that the solution to the pluriclosed flow is non-collapsing. And we prove that pluriclosed solitons with torsion potential are in fact Kähler Ricci solitons.

1 Introduction

The pluriclosed flow is introduced by Streets and Tian in [1]. Motivated by the celebrated work of Perelman on Ricci flow [2–4], they consider using a parabolic flow of Hermitian metrics to study the classification of Kodaira's VII surface (e.g. see [5, 6]). The Kähler Ricci flow does not work here since VII surfaces are non-Kähler. The pluriclosed flow preserves the pluriclosed condition, i.e. $\partial\bar{\partial}\omega = 0$. And from a theorem of Gauduchon [7], we can always find a pluriclosed metric ω, on compact complex surface. So it is a powerful tool for studying the VII surface.

One special case is the flow beginning with a Hermitian-symplectic form. On a complex manifold, a Hermitian-symplectic form is a real closed 2 form whose (1,1)-part is positive. This notion is introduced by Streets and Tian and in dimension 2 they prove that only Kähler surfaces admit Hermitian-symplectic forms [1]. It is easy to see that the (1,1)-part is a pluriclosed metric, and Ye [8] proves that pluriclosed flow preserves the Hermitian-symplectic condition. Under the Hermitian-symplection assumption, we can find a monotonic quantity along this flow. And as an application, it is easy to see that in this case, the pluriclosed flow is non-collapsing using a result of Garcia-Fernandez and Streets [10]. Moreover, we can prove that if a pluriclosed flow soliton is Hermitian-symplectic, then it is a Kähler Ricci soliton.

Actually, all those properties above are valid for a more general case, pluriclosed metric with torsion potential, which means there exists a (2,0)-form ϕ such that $\partial \omega + \bar{\partial} \phi = 0$. In dimension 2, pluriclosed metric with torsion potential is equivalent to the Hermitian-symplectic condition.

2 Notions and notations

In this section, we recall some basic notions and facts about pluriclosed flow.

Lemma 1 *On a Hermitian manifold (M^{2n}, J, g), there exists a unique Hermitian connection such that its torsion tensor is a 3-form.*

Definition 2 A family of Hermitian metrics $\omega(t)$ satisfies the pluriclosed flow if

$$\partial_t \omega(t) = -\rho^{1,1}(\omega), \quad \omega(0) = \omega_0.$$

Here $\rho^{1,1}(\omega)$ is the (1,1)-part of Bismut Ricci form and ω_0 is a pluriclosed metric.

Definition 3 A pluriclosed metric ω has torsion potential if there exists a (2,0)-form ϕ such that

$$\partial \omega + \bar{\partial} \phi = 0.$$

Lemma 4 *The pluriclosed flow preserves torsion potential condition.*

Proof Notice that a pluriclosed metric with torsion potential is equivalent to $[\partial \omega] = 0 \in \mathrm{H}_{\bar{\partial}}^{2,1}(M, \mathbb{C})$. Since the Bismut Ricci form ρ is real and closed, we obtain the evolution equation of torsion

$$\partial_t \partial \omega = -\partial \rho^{1,1} = \bar{\partial} \rho^{2,0}.$$

It is easy to see that the class $[\partial\omega] \in \mathrm{H}^{2,1}_{\bar{\partial}}(M,\mathbb{C})$ do not change along this flow. □

Definition 5 A metric is the soliton of pluriclosed flow if there exists a real constant λ and a vector field X such that

$$\rho^{1,1} = \lambda\omega + \mathcal{L}_X\omega$$

We call the soliton expanding, shrinking and steady if $\lambda < 0, \lambda = 0, \lambda > 0$, respectively. The most simple case of soliton is the case that $X = 0$. In this case, we also call the metric is Bismut Einstein with Einstein constant λ. For Bismut Einstein metrics, we know that if $\lambda \neq 0$, then the metric is Kähler and if $\lambda = 0$, then it is Bismut Ricci flat [9].

We will use ∇ to denote the Chern connection and use Δ to denote the Laplacian with respect to Chern connection. For convenience, sometimes we will omit the contraction taking by metric. For example, we will write the Chern Laplacian as

$$\Delta = g^{\bar{q}p}\nabla_p\nabla_{\bar{q}} = \nabla_a\nabla_{\bar{a}}$$

Here we use the same subscripts to denote the contraction taking by metric.

And we will use Ω to denote the curvature operator of Chern connection and use T to denote the torsion tensor. Then we can rewrite the pluriclosed flow as

$$\partial_t g = -S + Q$$

where $S_{i\bar{j}} = \Omega_{a\bar{a}i\bar{j}}$ and $Q_{i\bar{j}} = T_{iab}T_{\bar{j}\bar{a}b}$.

3 Main results

Given a pluriclosed metric ω_0 with torsion potential. We can find a (2,0)-form ϕ_0 satisfying $\partial\omega_0 + \bar{\partial}\phi_0 = 0$. In addition, if we run pluriclosed flow beginning with ω_0, then we can add an extra equation to the pluriclosed flow.

$$\partial_t\omega = -\rho^{1,1}, \quad \partial_t\phi = -\rho^{2,0}, \quad \omega(0) = \omega_0, \quad \phi(0) = \phi_0$$

Under the torsion potential assumption, we can reformulate the evolution equation of ϕ.

Lemma 6 *If the initial data is a pluriclosed metric with torsion potential, then the evolution equation of ϕ is*

$$(\partial_t - \Delta)\phi = 0.$$

Proof By direct computation, we have

$$\partial_t(\partial\omega + \bar{\partial}\phi) = -\partial\rho^{1,1} - \bar{\partial}\rho^{2,0} = 0$$

So we have $\partial\omega(t) + \bar{\partial}\phi(t) = 0$. From [9, 10], we know that

$$-\rho_{ij}^{2,0} = -\nabla_a T_{ij\bar{a}}$$

Combining with the torsion potential condition, we have

$$-\rho^{2,0} = \Delta\phi$$

Thus we have

$$(\partial_t - \Delta)\phi = 0.$$

□

By direct computation (or see [9]), we have the next evolution equation of $|\phi|^2$.

Lemma 7 *In torsion potential case, we have*

$$(\partial_t - \Delta)|\phi|^2 = -|T|^2 - |\nabla\phi|^2 - \mathrm{tr}_{g,Q}(\phi \otimes \bar{\phi})$$

Here

$$\mathrm{tr}_{g,Q}(\phi \otimes \bar{\phi}) = \phi_{ij}\phi_{\bar{a}\bar{j}}Q_{a\bar{i}}$$

is a non-negative number since Q is non-negative by definition.

Applying the strong maximum principle, we have the next corollary.

Corollary 8 *Let $(g(t), \phi(t))$ be a solution to pluriclosed flow with torsion potential. Then $F(t) = \max\limits_{x \in M} |\phi(x,t)|^2$ is decreasing along the flow. In additional, $F(t)$ is strictly decreasing when the initial data is non-Kähler .*

Combining the modified entropy given in the book [10], we know that

Theorem 9 *The pluriclosed flow with torsion potential is non-collapsing.*

For solitons with torsion potential, we have

Theorem 10 *If a pluriclosed soliton ω satisfies $[\partial\omega] = 0 \in H_{\bar\partial}^{2,1}(M, \mathbb{C})$, then we have $\partial\omega = 0$.*

Proof We normalize the flow by

$$\partial_t\omega = -\rho^{1,1} + \lambda\omega, \quad \partial_t\phi = -\rho^{2,0} + \lambda\phi$$

Up to a normalization, the soliton of original flow is always the "steady" soliton of normalized flow. But by a direct computation, we can find that the evolution equation of $|\tilde\phi|^2$ is invariant.

$$(\partial_t - \Delta)|\tilde\phi|^2 \leqslant -|\tilde T|^2$$

Then by the definition of steady soliton, there exists a time-independent vector field X such that

$$X|\tilde\phi|^2 - \Delta|\tilde\phi|^2 \leqslant -|\tilde T|^2$$

The result concludes by the strong maximum principle. □

References

[1] J. Streets and G. Tian. *A parabolic flow of pluriclosed metrics*, Int. Math. Res. Not. IMRN, 2010.

[2] G. Perelman. *The entropy formula for the Ricci flow and its geometric applications*, arXiv math/0211159, 2002.

[3] G. Perelman. *Ricci flow with surgery on three-manifolds*, arXiv math/0303109, 2003.

[4] G. Perelman. *Finite extinction time for the solutions to the Ricci flow on certain three-manifolds*, arXiv math/0307245, 2003.

[5] W. P. Barth, K. Hulek, C. A. M. Peters, et al. *Compact Complex Surfaces*, Ergebnisse der Mathematik und ihrer Grenzgebiete. 3. Folge. A Series of Modern Surveys in Mathematics. New York: Springer, 2004.

[6] J. Streets. *Pluriclosed flow and the geometrization of complex surfaces*, The Geometric Analysis Conference—in honor of Gang Tian's 60th birthday, 2020.

[7] P. Gauduchon. *Le théorème de l'excentricité nulle*, C. R. Acad. Sci. Paris Sér. A-B, 1977.

[8] Y. N. Ye. *Pluriclosed flow and Hermitian-symplectic structures*, arXiv 2207.12643, 2022.

[9] Y. N. Ye. *Bismut Einstein metrics on compact complex manifolds*, arXiv 2212.04060, 2023.

[10] M. Garcia-Fernandez, J. Streets. *Generalized Ricci flow*, University Lecture Series, 76. Providence: American Mathematical Society, 2021.

Weyl's Lemma on $RCD(K, N)$-spaces and an Application to PDEs on $\mathbb{R}^n$

PENG YU, HUICHUN ZHANG, XIPING ZHU

Department of Mathematics, Sun Yat-Sen University, Guangzhou, China

E-mail: pengy86@mail2.sysu.edu.cn, zhanghc3@mail.sysu.edu.cn, stszxp@mail.sysu.edu.cn

Abstract This is a survey about Weyl's lemma on the setting of $RCD(K, N)$-spaces, and its applications to the L^1-Liouville property and the regularity theory of partial differential equations on Euclidean spaces.

1 Introduction

Consider the Laplace equations on $\mathbb{R}^n$

$$\Delta u = \mathrm{div}(\nabla u) = 0, \qquad \text{on} \quad \Omega,$$

where $\Omega \subset \mathbb{R}^n$ is a domain. One of its classical solutions (C^2-solution) is called a harmonic function; a $W^{1,2}_{\mathrm{loc}}(\Omega)$ weak solution is called a *weakly harmonic* function; and then a $L^1_{\mathrm{loc}}(\Omega)$ very weak solution in the following sense is called a *very weakly harmonic* function: u is in $L^1_{\mathrm{loc}}(\Omega)$ and satisfies

$$\int_\Omega u \Delta \eta \, dx = 0, \qquad \forall\, \eta \in C_0^\infty(\Omega).$$

Maybe the first result in the regularity theory is the Weyl's lemma, given by H. Weyl in 1940 ([40]). He proved that any very weakly harmonic function in $\mathbb{R}^n$ must

The second author was partially supported by NSFC 12025109. The third author was partially supported by NSFC 12271530.

be smooth. Extensions of Weyl's lemma are widely studied (see, for examples, [5, 17, 21, 35]). Recently, Di Fratta-Fiorenza ([15]) gave a short proof for the regularity of very weak solutions of Poisson equations via the Weyl's lemma.

In this paper, we will survey an extension of Weyl's lemma to the setting of $RCD(K, N)$-spaces and introduce some applications to partial differential equations on $\mathbb{R}^n$. The class of $RCD(K, N)$-spaces is the collection of metric measure spaces with a synthetic notion of "Ricci curvature $\geqslant K$ and dimension $\leqslant N$". The RCD theory has developed in [2–4, 18, 19, 28, 38, 39]. We refer to the survey [1] for the improvements of geometric analysis on $RCD(K, N)$-spaces. Important examples of $RCD(K, N)$-spaces include Ricci limit spaces ([9–11]) and finite-dimensional Alexandrov spaces with curvature bounded from below with respect to their Hausdorff measures ([33, 41]).

In Section 2, we will introduce some basic ingredients on $RCD(K, N)$-spaces. In Section 3, we will deal with the concept of distributions in $RCD(K, N)$-spaces, to make the very weakly harmonic functions to be well-defined in $RCD(K, N)$-spaces, and then we introduce the Weyl's lemma (Theorem 13) and give a sketch of its proof. As an application of these results, in Section 4, we establish an L^1-Liouville property for sub-harmonic functions on $RCD(K, N)$-spaces. Finally, we discuss how to apply this result to the classical regularity theory of PDE on $\mathbb{R}^n$, including a point-wise gradient estimates of solutions to elliptic equations of divergence form and some $W^{1,p}$-estimates for elliptic and parabolic equations on $\mathbb{R}^n$.

2   Preliminaries

Let $(X, d, \mathfrak{m})$ be a metric measure space, i.e. (X, d) is a complete and separable metric space endowed with a non-negative Borel metric which is finite on bounded sets. We denote by $B_r(x)$ the ball with radius r and center at $x \in X$. $L^p(X) := L^p(X, \mathfrak{m})$.

2.1 $RCD(K, N)$-spaces

Given a function $f \in C(X)$, the *pointwise Lipschitz constant* ([8]) of f at x is defined by

$$\mathrm{Lip} f(x) := \limsup_{y \to x} \frac{|f(y) - f(x)|}{d(x, y)} = \limsup_{r \to 0} \sup_{d(x,y) \leqslant r} \frac{|f(y) - f(x)|}{r},$$

where we put $\text{Lip} f(x) = 0$ if x is isolated. Clearly, $\text{Lip} f$ is a $\mathfrak{m}$-measurable function on X. The *Cheeger energy*, denoted by $\text{Ch}: L^2(X) \to [0, \infty]$, is defined ([2]) by

$$\text{Ch}(f) := \inf \Big\{ \liminf_{j \to \infty} \frac{1}{2} \int_X (\text{Lip} f_j)^2 d\mathfrak{m} \Big\},$$

where the infimum is taken over all sequences of Lipschitz functions $(f_j)_{j \in \mathbb{N}}$ converging to f in $L^2(X)$. In general, Ch is a convex and lower semi-continuous functional on $L^2(X)$.

Definition 1 A metric measure space $(X, d, \mathfrak{m})$ is called *infinitesimally Hilbertian* if the associated Cheeger energy Ch is quadratic.

Given an infinitesimally Hilbertian metric measure space $(X, d, \mathfrak{m})$, the energy $\mathcal{E} := 2\text{Ch}$ gives a canonical Dirichlet form on $L^2(X)$ with the domain $W^{1,2}(X) := D(\text{Ch})$. Let $(\Delta, D(\Delta))$ and $(H_t f)_{t \geqslant 0}$ denote the infinitesimal generator and the heat flow induced from $(\mathcal{E}, W^{1,2}(X))$. For any $f \in W^{1,2}(X)$, we denote by $|\nabla f| \in L^2(X)$ the minimal weak upper gradient of f (see for example [20, Sect. 2.1]). If $f \in W^{1,2}(X)$ is Lipschitz continuous, it is well-known that $|\nabla f| = \text{Lip} f$ $\mathfrak{m}$-a.e.

Definition 2 Let $K \in \mathbb{R}$ and $N \geqslant 1$. An infinitesimally Hilbertian metric measure space $(X, d, \mathfrak{m})$ is called an $RCD(K, N)$-*space* if the following conditions are satisfied:

(1) There exists $C > 0$ such that $\mathfrak{m}(B_r(x)) \leqslant C \exp(Cr^2)$;

(2) The *Bakry-Emery condition* $BE(K, N)$ holds: If $f \in D(\Delta)$ with $\Delta f \in W^{1,2}(X)$ and if $\phi \in D(\Delta) \cap L^\infty(X)$ with $\phi \geqslant 0$ and $\Delta \phi \in L^\infty(X)$, then we have

$$\frac{1}{2} \int_X \Delta \phi \cdot |\nabla f|^2 d\mathfrak{m} \geqslant \frac{1}{N} \int_X \phi \big((\Delta f)^2 \langle \nabla(\Delta f), \nabla f \rangle + K |\nabla f|^2 \big) d\mathfrak{m}.$$

(3) If $f \in W^{1,2}(X) \cap L^\infty(X)$ and $|\nabla f| \leqslant 1$, then it has a Lipschitz continuous representative.

Throughout this paper, we always assume that $(X, d, \mathfrak{m})$ is an $RCD(K, N)$-space with $N \in [1, +\infty)$ and $K \in \mathbb{R}$.

For any open domain $\Omega \subset X$, the space $W_0^{1,2}(\Omega)$ is the closure of $Lip_c(\Omega)$ with support in Ω under $W^{1,2}(X)$-norm. The space $W_{\text{loc}}^{1,2}(\Omega)$ is the space of

Borel functions $f : \Omega \to \mathbb{R}$ such that $f\chi \in W^{1,2}_{\text{loc}}(X)$ for any Lipschitz function $\chi : X \to [0, 1]$ such that $d(\text{supp}(\chi), X \setminus \Omega) > 0$, where a function $f\chi$ is taken 0 by definition on $X \setminus \Omega$.

For any $f, h \in W^{1,2}(X)$, the inner product $\langle \nabla f, \nabla h \rangle$ is given by

$$\langle \nabla f, \nabla h \rangle = \frac{1}{4} \left(|\nabla (f + h)|^2 - |\nabla (f - h)|^2 \right) \in L^1(X).$$

The inner product $\langle \nabla f, \nabla h \rangle$ provides the *Carré du champ* of the Dirichlet form $(\mathscr{E}, W^{1,2}(X))$. It is symmetric and bi-linear, and satisfies Cauchy-Schwarz inequality, Chain rule and Leibniz rule (see [36, Appendix] or [20]).

According to [3,36,37], there exists a heat kernel $p_t(x, y)$ on $(0, +\infty) \times X \times X$ such that

$$P_t f(x) = \int_X p_t(x, y) f(y) d\mathfrak{m}(y), \quad \forall\, t > 0$$

for every $f \in L^2(X)$. It was proved ([3,37]) that $(x, y) \mapsto p_t(x, y)$ is locally Hölder continuous on $X \times X$ for each $t > 0$, and that $t \mapsto p_t(x, y)$ is analytic in $(0, +\infty)$ for each $x, y \in X$. Moreover, it was shown in [23] that there exist constants $C_1, C_2 > 0$ depending only on N and K such that

$$\frac{1}{C_1 \mathfrak{m} \left(B_{\sqrt{t}}(x) \right)} \exp \left\{ -\frac{d^2(x, y)}{3t} - C_2 t \right\} \leqslant p_t(x, y)$$

$$\leqslant \frac{C_1}{\mathfrak{m} \left(B_{\sqrt{t}}(x) \right)} \exp \left\{ -\frac{d^2(x, y)}{5t} + C_2 t \right\} \tag{2.1}$$

for all $t > 0$ and all $x, y \in X$,

$$|\nabla p_t(\cdot, y)| (x) \leqslant \frac{C_1}{\sqrt{t} \cdot \mathfrak{m} \left(B_{\sqrt{t}}(x) \right)} \exp \left\{ -\frac{d^2(x, y)}{5t} + C_2 t \right\} \tag{2.2}$$

for all $t > 0$ and $\mathfrak{m}$-a.e. $x, y \in X$, and

$$\left| \frac{\partial}{\partial t} p_t(x, y) \right| = |\Delta p_t(\cdot, y)| (x) \leqslant \frac{C_1}{t \cdot \mathfrak{m} \left(B_{\sqrt{t}}(x) \right)} \exp \left\{ -\frac{d^2(x, y)}{5t} + C_2 t \right\} \tag{2.3}$$

holds for all $t > 0$ and $\mathfrak{m}$-a.e. $x, y \in X$. An important property of $p_t(x, y)$ is the stochastic completeness (see [36]):

$$\int_X p_t(x, y) d\mathfrak{m}(y) = 1, \quad \forall t > 0 \text{ and } x \in X. \tag{2.4}$$

2.2 Test functions and "good" cut-off functions

The class of *test functions* on $RCD(K, N)$-spaces was introduced in [20]:

$$\text{Test}^\infty(X) := \left\{ f \in D(\Delta) \cap L^\infty(X) : |\nabla f| \in L^\infty(X), \ \Delta f \in L^\infty(X) \cap W^{1,2}(X) \right\},$$

which is an algebra and is dense in $W^{1,2}(X)$ (see [20, Sect.6.1.3]). Given any $\phi \in \text{Test}^\infty(X)$, it is well-known ([4]) that it has a (Lipschitz) continuous representative. Hence we always assume that every function in $\text{Test}^\infty(X)$ is continuous. We equip $\text{Test}^\infty(X)$ with the norm

$$\|\phi\|_E := \|\phi\|_{L^\infty(X)} + \||\nabla \phi|\|_{L^\infty(X)} + \|\Delta \phi\|_{L^\infty(X)},$$

and denote by $E(X)$ the linear normed space $(\text{Test}^\infty(X), \|\cdot\|_E)$.

Remark 3 The class of test functions plays the role of the class of C^∞-functions in smooth manifolds. Of course, one can consider the collection of functions $f \in D(\Delta^j)$ for each $j \in \mathbb{N}$ as another class of test functions.

The following lemma is a standard fact about the heat kernel.

Lemma 4 (1) *For any $t > 0$ and $x_0 \in X$, the function $p_t(x_0, \cdot) \in \text{Test}^\infty(X)$.*
(2) *For any $\phi \in \text{Test}^\infty(X)$ and $t > 0$, it holds $P_t \phi \in \text{Test}^\infty(X)$.*

The existence of good cut-off functions is an important ingredient in geometric analysis. For an n-dimensional Riemannian manifold M^n with Ricci curvature bounded from below, one can estimate merely the upper bounds for the Laplacian of distance functions, by the Laplace comparison theorem. In this aspect, Schoen-Yau in [34, Theorem 4.2 in Chapter I] constructed distance-like functions on M^n having L^∞-bounds of the Laplacian. Cheeger-Colding [9] found good cut-off functions with L^∞ estimate on the Laplacian, which is a key technical tool in the Cheeger-Colding theory of Ricci limit spaces (see [9–11]). The existence of a regular cut-off function in RCD setting was proved in [4, 22, 30]. We need the following variant of [30, Lemma 3.1].

Lemma 5 *Let $(X, d, \mathfrak{m})$ be an $RCD(K, N)$-space with $K \in \mathbb{R}$ and $1 \leqslant N < \infty$. For each $r_0 > 0$, there exists a constant $C_{K,N,r_0} > 0$ such that for any ball $B_R(x) \subset X$ with radius $R \geqslant r_0$, there exists a cut-off function $\eta : X \to [0,1]$ such that*

(i) $\eta \equiv 1$ *on* $B_R(x)$ *and* $\text{supp}(\eta) \subset B_{2R}(x)$;

(ii) $\eta \in \text{Test}^\infty(X)$, and moreover for $\mathfrak{m}$-a.e. $x \in X$,

$$|\Delta\eta|(x) + |\nabla\eta|(x) \leqslant \frac{C_{N,K,r_0}}{R}.$$

(Remark that the constant C_{N,K,r_0} does not depend on R.)

Remark 6   If $K = 0$, then by a rescaling argument, one can get a cut-off function such that

$$|\Delta\eta| + |\nabla\eta|^2 \leqslant \frac{C_N}{R^2}.$$

If $K \neq 0$, it is natural to expect the existence of a cut-off function satisfying

$$|\Delta\eta| + |\nabla\eta|^2 \leqslant \frac{C_{K,N,r_0}}{R^2}.$$

We thank Professor Miaomiao Zhu for pointing out this problem.

We recall the following measure-valued Laplacian introduced in [19].

Definition 7   Let $\Omega \subset X$ be an open subset and let $f \in W^{1,2}_{\text{loc}}(\Omega)$. It is called $f \in D(\Delta)$ on Ω if there exists a signed Radon measure μ on Ω such that

$$\int_\Omega \phi \, d\mu = -\int_\Omega \langle \nabla\phi, \nabla f \rangle d\mathfrak{m}, \quad \forall \phi \in Lip_c(\Omega).$$

Such μ is unique and is denoted by Δf (which depends on Ω).

Both the Chain rule and the Leibniz rule for this Δ were proved in [19, Chapter 4]. When $\Omega = X$, it was showed ([19, Proposition 4.24]) that for any $f \in W^{1,2}(X)$,

$$f \in D(\Delta) \quad \Longleftrightarrow \quad f \in D(\ ) \text{ and } f = g \cdot \mathfrak{m} \text{ for some } g \in L^2(X).$$

In this case $f = \Delta f \cdot \mathfrak{m}$.

3   Weyl's Lemma on $RCD(K, N)$-spaces

Let $(X, d, \mathfrak{m})$ be an $RCD(K, N)$-space with $K \in \mathbb{R}$ and $N \in [1, +\infty)$.

3.1　Distributional Laplacian and Weyl's Lemma

Let $\Omega \subset X$ be an open domain, we set

$$\mathrm{Test}_c^\infty(\Omega) := \big\{\phi \in \mathrm{Test}^\infty(X)\big|\ \mathrm{supp}(\phi) \subset \Omega \text{ and } \mathrm{supp}(\phi) \text{ is compact}\big\}$$

and　　$\mathrm{Test}_{c,+}^\infty(\Omega) := \{\phi \in \mathrm{Test}_c^\infty(\Omega)|\ \phi \geqslant 0\}.$

From Lemma 5, for each open domain $\Omega \subset X$ and every compact set $Q \subset \Omega$, there exists $\eta \in \mathrm{Test}_c^\infty(\Omega)$ such that $\eta \equiv 1$ on Q and $0 \leqslant \eta \leqslant 1$. The space $\mathrm{Test}_{c,+}^\infty(\Omega)$ is dense in $\{f \in Lip_c(\Omega)|f \geqslant 0\}$. Consequently, the space $\mathrm{Test}_c^\infty(\Omega)$ is dense in $W_0^{1,2}(\Omega)$.

Definition 8　Let $\Omega \subset X$. A *distribution* T in Ω is a linear functional on $\mathrm{Test}_c^\infty(\Omega)$ such that for every compact set $Q \subset \Omega$ there exists a constant C_Q such that

$$|T(\phi)| \leqslant C_Q\|\phi\|_E, \quad \forall \phi \in \mathrm{Test}_c^\infty(\Omega) \text{ with } \mathrm{supp}(\phi) \subset Q.$$

It is clear that every element $f \in E'(X)$, the dual space of $E(X)$, defines a distribution on X.

Definition 9　Let $f \in L_{\mathrm{loc}}^1(\Omega)$, the *distributional Laplacian* of f, denoted by $\Delta_{\mathscr{D},\Omega}f$, is defined as a linear functional on $\mathrm{Test}_c^\infty(\Omega)$ by

$$\Delta_{\mathscr{D},\Omega}f(\phi) := \int_\Omega f\Delta\phi\,\mathrm{d}\mathfrak{m}, \quad \forall \phi \in \mathrm{Test}_c^\infty(\Omega).$$

If $\Omega = X$, we denote $\Delta_{\mathscr{D}}f := \Delta_{\mathscr{D},X}f$ for any $f \in L_{\mathrm{loc}}^1(X)$.

For each $f \in L_{\mathrm{loc}}^1(\Omega)$, it is easy to check that $\Delta_{\mathscr{D},\Omega}f$ is a distribution in Ω. Given two domains $\Omega \subset \Omega'$ and a function $f \in L_{\mathrm{loc}}^1(\Omega')$, it is not hard to check that $\Delta_{\mathscr{D},\Omega}f = \Delta_{\mathscr{D},\Omega'}f$ in Ω.

Definition 10　Let $\Omega \subset X$ be an open domain. A function $u \in L_{\mathrm{loc}}^1(\Omega)$ is called a *very weak subsolution (resp. supersolution)* of Poisson equation $\Delta_{\mathscr{D},\Omega}u = f$ on Ω for some $f \in L_{\mathrm{loc}}^1(\Omega)$ if

$$\int_\Omega u\Delta\phi\,\mathrm{d}\mathfrak{m} \geqslant \text{ (resp. } \leqslant) \int_\Omega f\phi\,\mathrm{d}\mathfrak{m}, \quad \forall \phi \in \mathrm{Test}_c^\infty(\Omega) \text{ and } \phi \geqslant 0. \tag{3.1}$$

A function $u \in L_{\mathrm{loc}}^1(\Omega)$ is called a *very weak solution* of $\Delta_{\mathscr{D},\Omega}u = f$ on Ω if it is both very weak subsolution and supersolution. If $f = 0$, such u is called very weakly harmonic subharmonic (resp. superharmonic, harmonic) on Ω.

It has been proved ([32]) that the distribution Laplacian is compatible with the measure-valued Laplacian in [19].

Lemma 11 *Let $\Omega \subset X$ be an open domain. If $f \in W^{1,2}_{\text{loc}}(\Omega)$, then $\Delta_{\mathscr{D},\Omega} f = \Delta f$ on $\text{Test}^\infty_c(\Omega)$. In particular, if u is subharmonic/superharmonic on Ω then it is very weakly subharmonic/superharmonic on Ω.*

Recall that P_t is a family of linear contractions on $L^2(X)$ for any $t > 0$, it can be uniquely extended to be a family of linear contractions on $L^p(X)$ for any $p \in [1, +\infty]$. We denote the infinitesimal generator of P_t on $L^p(X)$ by $\Delta^{(p)}$. We have also the compatibility of $\Delta^{(p)}$ and $\Delta_{\mathscr{D},X}$.

Lemma 12 *For each $p \in [1, \infty)$, if $f \in D(\Delta^{(p)})$, then $\Delta_{\mathscr{D}} f = \Delta^{(p)} f$ in the sense of distributions on X. Here $\Delta^{(p)}$ is the generator of P_t as semi-group acting on $L^p(X)$. That is,*

$$D(\Delta^{(p)}) := \left\{ f \in L^p(X) \,\middle|\, \lim_{t \to 0^+} \frac{P_t f - f}{t} \text{ exists in } L^p \right\} \quad \text{and}$$

$$\Delta^{(p)} f := \lim_{t \to 0^+} \frac{P_t f - f}{t}.$$

The main result of this paper is the following Weyl's lemma on RCD-setting.

Theorem 13 (Weyl's lemma) *Let $(X, d, \mathfrak{m})$ be an $RCD(K, N)$-space for some $1 \leqslant N < \infty$ and $K \in \mathbb{R}$. Let $\Omega \subset X$ be an open domain. If $u \in L^1_{\text{loc}}(\Omega)$ is a very weakly harmonic function on Ω, then u is locally Lipschitz continuous on Ω (i.e. it has a locally Lipschitz continuous representative). In particular, it is in $W^{1,2}_{\text{loc}}(\Omega)$.*

3.2 The sketch of the proof of Theorem 13

The proof is divided as the following two steps:

Step 1: It is shown in [32] that

Lemma 14 *Let u and f be in $L^1(X)$. Suppose that u is a very weak subsolution of Poisson equation $\Delta_{\mathscr{D}} u = f$ on X. Then*

$$\int_X u \Delta h \, d\mathfrak{m} \geqslant \int_X f h \, d\mathfrak{m}, \qquad \forall h \in \text{Test}^\infty(X) \text{ and } h \geqslant 0. \tag{3.2}$$

This lemma makes that it is possible for us to use the $P_t \phi$ as a test function.

Step 2: Given a function $v \in L^1(X)$ such that $v(x) = 0$ a.e. $x \in X \setminus B_R(x_0)$ for some ball $B_R(x_0) \subset X$. Suppose that $\Delta_{\mathscr{D}} v(\phi) = 0$ for any nonnegative $\phi \in \text{Test}_c^\infty(X)$ with $\text{supp}(\phi) \subset B_{R/2}(x_0)$. Let $p_t(x, y)$ be the heat kernel on X. For any $t > 0$, we set

$$v_t(x) := P_t v(x) = \int_X p_t(x, y) v(y) d\mathfrak{m}, \quad \forall t > 0.$$

Lemma 15 (Key estimate) *There exists a constant $c > 0$ such that for any $t \in (0, R^2)$ it holds*

$$\int_{B_{R/4}(x_0)} \langle \nabla v_t, \nabla \phi \rangle d\mathfrak{m} \leqslant c \int_{B_{R/4}(x_0)} |\phi| d\mathfrak{m}, \quad \forall \phi \in Lip_c\big(B_{R/4}(x_0)\big). \quad (3.3)$$

The first application of Theorem 13 is the following local regularity result for the very weak solutions of Poisson equations on RCD-spaces.

Corollary 16 *Suppose $f \in L^2_{\text{loc}}(\Omega)$. If $u \in L^1_{\text{loc}}(\Omega)$ is a very weak solution of Poisson equation on Ω in the sense of*

$$\int_\Omega u \Delta \phi d\mathfrak{m} = \int_\Omega f \phi d\mathfrak{m}, \quad \forall \phi \in \text{Test}_c^\infty(\Omega) \text{ and } \phi \geqslant 0. \quad (3.4)$$

Then u is in $W^{1,2}_{\text{loc}}(\Omega)$.

4　The Liouville property for L^1 (sub)harmonic functions

In this section, we will prove the Liouville-type theorem for L^1 subharmonic/harmonic functions on $RCD(K, N)$-spaces, which is an extension of Peter Li's result in [25]. Let $(X, d, \mathfrak{m})$ be an $RCD(K, N)$-space with $K \in \mathbb{R}$ and $N \in [1, +\infty)$.

Corollary 17 *Let $(X, d, \mathfrak{m})$ be an $RCD(K, N)$-space with $K \in \mathbb{R}$ and $1 \leqslant N < \infty$, then any $L^1(X)$ very weakly subharmonic function on X must be identically constant (namely, it has a constant representative).*

Recall that Li-Schoen [26] and Chung [12] gave some examples that a complete Riemannian manifold admits non-constant nonnegative L^1 harmonic functions. Under an assumption of Ricci lower bounds, P. Li [25] proved a Liouville-type result for L^1 subharmonic functions on complete Riemannian manifolds. We will prove Corollary 17 along the same line in [25] and combine Theorem 13.

5   Applications to PDEs in $\mathbb{R}^n$

Let $n \geqslant 2$ and let $D \subset \mathbb{R}^n$ be a bounded domain.

5.1   Point-wise gradient estimates for solutions of elliptic equations

We will give a point-wise gradient estimate for weak solutions of a class of elliptic equations on Euclidean spaces with *dis-continuous coefficients*.

Suppose that $u \in W^{1,2}_{\text{loc}}(D)$ is a weak solution of elliptic equations of divergence form

$$Lu := \text{div}(A(x)Du) = \sum_{i,j=1}^{n} \partial_i(a^{ij}\partial_j u) = 0, \quad 1 \leqslant i, j \leqslant n, \qquad (5.1)$$

where the matrix of coefficients $A(x) = (a^{ij})_{i,j=1}^{n}$ is symmetric, bounded, measurable and uniformly elliptic, i.e.

$$\|a^{ij}\|_{L^\infty(D)} \leqslant \Lambda, \quad \lambda|\xi|^2 \leqslant \sum_{i,j=1}^{n} a^{ij}(x)\xi_i\xi_j, \quad \text{a.e. } x \in D, \quad \forall \xi \in \mathbb{R}^n \qquad (5.2)$$

for some $\lambda, \Lambda > 0$. It is well known [5] that any weak solution u of $Lu = 0$ is in $C^1(D)$ provided that A satisfies the Dini continuity condition. Recently, this Dini continuity condition was weaken to other Dini-type conditions by Y. Li [27], Dong-Kim [16] and Maz'ya-McOwen [29]. However, an example constructed by Jin-Maz'ya-Schaftingen [24] shows that, general speaking, the continuity of the coefficient is not sufficient to ensure $u \in \text{Lip}_{\text{loc}}(D)$.

Inspired by the regularity of harmonic functions on RCD-spaces, we first introduce a class of models for elliptic coefficient, which plays the role of constant coefficient in the classical regularity theory.

Definition 18   Let $\overline{A} = (\overline{a}^{ij}) : \mathbb{R}^n \to \mathbb{R}^n \times \mathbb{R}^n$ be symmetric, bounded, measurable and uniformly elliptic on the each ball $B_R(0) \subset \mathbb{R}^n$ with $R > 0$.

Let $n \geqslant 3$. We first introduce a Riemannian metric $\overline{g}_{ij}(x)$ on $\mathbb{R}^n$ by

$$(\overline{g}_{ij})^{-1}(x) = \overline{g}^{ij}(x) := \frac{\overline{a}^{ij}(x)}{\left[\det(\overline{a}^{ij}(x))\right]^{\frac{1}{n-2}}}, \qquad 1 \leqslant i, j \leqslant n. \qquad (5.3)$$

We denote by $d_{\overline{g}}$ the distance function induced by Riemannian metric $(\mathbb{R}^n, \overline{g}_{ij})$.

The coefficient $\overline{A}$ is called a *conical coefficient of nonnegative curvature* if the metric space $(\mathbb{R}^n, d_{\overline{g}})$ is a cone with nonnegative curvature in the sense of Alexandrov.

We remark that if $\overline{a}^{ij}$ is constant coefficient, then the Riemannian metric $\overline{g}_{ij} = const$, and then $(\mathbb{R}^n, \overline{g}_{ij})$ is flat (i.e., sectional curvature is zero).

There are also some *dis-continuous* conical coefficient of nonnegative curvature. For example, consider the graph Γ_f of a convex function

$$f(x_1, x_2, \cdots, x_n) = \left(a_1 x_1^2 + a_2 x_2^2 + \cdots + a_n x_n^2\right)^{1/2},$$

where $a_i \in (0, +\infty)$ for all $i = 1, \cdots, n$. It is well known that Γ_f is a cone of nonnegative curvature in the sense of Alexandrov. Under the coordinate system $(x_1, x_2, \cdots, x_n)$, its Riemannian metric is given by

$$\overline{g}_{ij} = \delta_{ij} + \frac{a_i x_i}{f} \cdot \frac{a_j x_j}{f},$$

for all $1 \leqslant i, j \leqslant n$. Then $\overline{A}(x) = (\overline{a}^{ij}(x))$ with

$$\overline{a}^{ij} := \sqrt{\det(\overline{g}_{ij})}\,\overline{g}^{ij} = \left(1 + \frac{\sum_{i=1}^n (a_i x_i)^2}{f^2}\right)^{1/2} \left(\delta_{ij} + \frac{a_i x_i}{f} \cdot \frac{a_j x_j}{f}\right)^{-1}$$

is a conical coefficient of nonnegative curvature, and $\overline{a}^{ij}(x)$ is not continuous at 0.

Theorem 19   *Let $u \in W_{\mathrm{loc}}^{1,2}(B_1(0))$ be a weak solution of elliptic equations of divergence form $\mathrm{div}(A(x)Du) = 0$, where the matrix of coefficients $A(x) = (a^{ij})_{i,j=1}^n$ is symmetric, bounded, measurable and uniformly elliptic. Let $n \geqslant 3$.*

Suppose that A is Dini asymptotic to a conical coefficient of nonnegative curvature in the sense that there exists a conical coefficient of nonnegative curvature $\overline{A} = (\overline{a}^{ij})$ such that

$$\int_0^1 \frac{\omega_{A,\overline{A}}(t)}{t} dt < \infty, \quad \text{where } \omega_{A,\overline{A}}(t) := \|A(x) - \overline{A}(x)\|_{L^\infty(B_t(0))}. \qquad (5.4)$$

Then

$$\limsup_{r \to 0^+} \fint_{B_r(0)} |\nabla u|^2 d\mathcal{L}^n \leqslant C_0 \fint_{B_1(0)} |\nabla u|^2 d\mathcal{L}^n, \tag{5.5}$$

where the constant C_0 depends only on n, λ, Λ and $\omega_{A,\overline{A}}$, and $\mathcal{L}^n$ is the n-dimensional Lebesgue measure on $\mathbb{R}^n$.

This extends the result in [5].

The idea of the proof is the method of "freezon the coefficients" by replacing the model from constant coefficients to the conical coefficients of nonnegative curvature, and using the following lemma.

Lemma 20   *Let $(C, d_c, \mathfrak{m}_c)$ be an $RCD(0, N)$ cone with vertex o, over $(\Sigma, d_\Sigma, \mathfrak{m}_\Sigma)$. Suppose that $u \in W^{1,2}(B_R(o))$ is a (weakly) harmonic function, then*

$$\fint_{B_{r_1}(o)} |\nabla u|^2 d\mathfrak{m}_c \leqslant \fint_{B_{r_2}(o)} |\nabla u|^2 d\mathfrak{m}_c \tag{5.6}$$

for all $0 < r_1 < r_2 < R$.

5.2 $W^{1,p}$-estimates

Let Ω be an open subset of $\mathbb{R}^n$, $n \geqslant 2$. We consider the elliptic equations in divergence form

$$Lu := -\sum_{ij} \partial_i(a_{ij}(x)\partial_j u(x)) = \mathrm{div} F(x) = \sum_i \partial_i f_i. \tag{5.7}$$

where the matrix of coefficients $A(x) = (a_{ij}(x))_{n \times n}$ as the above subsection. We say that $u \in W^{1,2}_{\mathrm{loc}}(\Omega)$ is a weak solution to (5.7) if

$$\int_\Omega \sum_{ij} a_{ij}(x)\partial_j u(x)\partial_i \phi(x)dx = -\int_\Omega F(x) \cdot \nabla \phi(x)dx \tag{5.8}$$

for all $\phi \in \mathrm{Lip}_0(\Omega)$, the space of Lipschitz functions on Ω with compact support.

We are concerned with the L^p-estimate for the gradient of weak solutions to (5.7) in the form

$$\int_{B_r(x)} |\nabla u|^p dx \leqslant C \int_{B_{2r}(x)} (|u|^p + |F|^p)dx. \tag{5.9}$$

This estimate (5.9) with constant coefficients was first proved by Calderón-Zygmund. It was extended to the case where the coefficients are continuous by

De Giorgi and Companatto, the case where the coefficients are in VMO space (hence, may be discontinuous) [5], and the case where the coefficients have small oscillation [7], or small BMO semi-norms [6]. Let us recall the concept of BMO semi-norms of matrix A,

$$\|A\|_{BMO(\Omega)} := \sup_{x \in \Omega} \sup_{0 < r < R} \fint_{B_r(x)} |A - \overline{A}_r|^2 dx, \tag{5.10}$$

where $\overline{A}_r = \fint_{B_r(x)} A(x) dx$.

Theorem 21 (Theorem 1.5 in [6]) *Let Ω be an open subset of $\mathbb{R}^n$, $n \geqslant 2$ and $2 \leqslant p < \infty$. There exists a number $\delta = \delta(p, \lambda, \Lambda, n) > 0$ such that for all A which is uniformly elliptic on Ω with elliptic constants λ, Λ (see (5.2)) and if u is a weak solution to (5.7) such that $\|A\|_{BMO(\Omega)} \leqslant \delta$ and $F \in L^p(\Omega, \mathbb{R}^n)$, then $u \in W^{1,p}_{\text{loc}}(\Omega)$ and (5.9) holds on any ball $B_r(x)$ with $B_{2r}(x) \subset \Omega$, where the constant C is independent of u and F.*

Before stating our main result, we recall the notation of semi-convex functions.

Definition 22 Let Ω be an open subset of $\mathbb{R}^n$ and $a \in \mathbb{R}$, we say that a function f is a-convex on Ω if $f(x) - \dfrac{a}{2}\|x\|^2$ is convex on Ω.

Remark 23 (1) If f is a-convex for some $a \in \mathbb{R}$, then $f \in \text{Lip}_{\text{loc}}(\Omega)$.

(2) If $f \in C^2$, then f is a-convex if and only if $\text{Hess} f \geqslant a \cdot I$, where I is the identity matrix.

In [31], we proved the following $W^{1,p}$-estimates:

Theorem 24 *Let Ω be an open subset of $\mathbb{R}^n$, $a \in \mathbb{R}$ and $2 \leqslant p < \infty$. There exists a number $\delta = \delta(p, n, \lambda, \Lambda, a, M) > 0$ such that for all A which is uniformly elliptic on Ω with elliptic constants λ, Λ and $F \in L^p(\Omega, \mathbb{R}^n)$ the following property holds:*

If $u \in W^{1,2}(\Omega)$ is a weak solution to (5.7) and if there exists an a-convex function Φ on Ω with $|\nabla \Phi| + |\nabla|\nabla \Phi|| \leqslant M$ on Ω such that

$$\|A - A_0\|_{L^\infty(\Omega)} \leqslant \delta, \tag{5.11}$$

then $u \in W^{1,p}_{\text{loc}}(\Omega)$ and the estimate (5.9) holds for all ball $B_r(x)$ with $B_{2r}(x) \subset \Omega$, where $A_0(x) = (a_{ij}^0(x))_{n \times n}$, is defined by

$$a_{ij}^0(x) = (1 + |\nabla\Phi(x)|^2)^{\frac{1}{2}} \left(\delta_{ij} - \frac{\partial_i\Phi(x)\partial_j\Phi(x)}{1 + |\nabla\Phi(x)|^2} \right) \quad a.e. \ x \in \Omega. \tag{5.12}$$

The parabolic case was also dealt with in [31]. Let $\Omega = B_2(0)$ be a bounded domain in $\mathbb{R}^n$. In this section, we consider the interior $W^{1,p}$ estimates for weak solution to parabolic equations in divergence form

$$u_t = \text{div}(a(x,t,\nabla u)) + \text{div}F, \quad x \in \Omega, \quad t > 0, \tag{5.13}$$

where $a : \Omega \times (0,\infty) \times \mathbb{R}^n \to \mathbb{R}^n$, $F \in L^p(\Omega \times (0,\infty), \mathbb{R}^n)$. We assume that $a(x,t,\xi)$ is a Caratheodory function (in the sense that is measurable in (x,t) and continuous with respect to ξ for each x) and satisfies the following conditions:

1. $a(x,t,0) = 0$;
2. $\langle a(x,t,\xi) - a(x,t,\eta), (\xi - \eta) \rangle \geqslant \gamma|\xi - \eta|^2$;
3. $|a(x,t,\xi)| \leqslant \Gamma|\xi|$;
4. a is linear about ξ;

where γ and Γ are positive constants.

In [31], we proved that

Theorem 25   *Given $n \geqslant 2$, $p > 2$, $\Omega = B_2(0)$. There exists a constant $\delta = \delta(n,p,\lambda,\Lambda,a,M)$ such that if $F \in L^p(R;\mathbb{R}^n)$ and if u is a weak solution to (5.13) with*

$$|a(x,t,\xi) - A_0(\xi)| \leqslant \delta|\xi| \tag{5.14}$$

uniformly in x and t. Then $|\nabla u| \in L_{loc}^q$ for all $2 < q < p$.

References

[1] L. Ambrosio, *Calculus, heat flow and curvature-dimension bounds in metric measure spaces*, Proceedings of the International Congress of Mathematicians—Rio de Janeiro 2018. Vol. I. Plenary lectures, Hackensack: World Sci. Publ., 2018, 301–340.

[2] L. Ambrosio, N. Gigli, and G. Savaré, *Calculus and heat flow in metric measure spaces and applications to spaces with Ricci bounds from below*, Invent. Math., **195** (2014), no. 2, 289–391.

[3] L. Ambrosio, N. Gigli, and G. Savaré, *Metric measure spaces with Riemannian Ricci curvature bounded from below*, Duke Math. J., **163** (2014), no. 7, 1405–1490.

[4] L. Ambrosio, A. Mondino and G. Savaré, *On the Bakry-Émery condition, the gradient estimates and the local-to-global property of* $\mathsf{RCD}^*(K, N)$ *metric measure spaces*, J. Geom. Anal., **26** (2016), no.1, 24–56.

[5] C. C. Burch, *The Dini condition and regularity of weak solutions of elliptic equations*, J. Diff. Eq., **30** (1978), 308–323.

[6] S. Byun, L. Wang, *Elliptic equations with BMO coefficients in Reifenberg domains.* Comm. Pure Appl. Math., **57** (2004), no. 10, 1283-1310.

[7] L. A. Caffarelli, I. Peral, *On* $W^{1,p}$ *estimates for elliptic equations in divergence form.* Comm. Pure Appl. Math., **51** (1998), no. 1, 1-21.

[8] J. Cheeger, *Differentiability of Lipschitz functions on metric measure spaces*, Geom. Funct. Anal., **9** (1999), no. 3, 428–517.

[9] J. Cheeger, T. Colding, *Lower bounds on Ricci curvature and the almost rigidity of warped products*, Ann. of Math., **144** (1996), 189–237.

[10] J. Cheeger, T. Colding, *On the structure of spaces with Ricci curvature bounded below I*, J. Differ. Geom., **46** (1997), 406–480.

[11] J. Cheeger, T. Colding, *On the structure of spaces with Ricci curvature bounded below II, III*, J. Differ. Geom., **54** (2000), 13–35; **54** (2000), 37–74.

[12] L. O. Chung, *Existence of harmonic* L^1 *functions in complete Riemannian manifolds*, Proc. Amer. Math. Soc., **88** (1983), 531–532.

[13] G. De Philippis and N. Gigli, *Non-collapsed spaces with Ricci curvature bounded from below*, J. Éc. polytech. Math., **5** (2018), 613–650.

[14] G. De Philippis and J. Núñez-Zimbrón, *The behavior of harmonic functions at singular points of RCD-spaces,* J. Geom. Anal., online and available at: https://doi.org/10.1007/s00229-021-01365-9.

[15] G. Di Fratta and A. Fiorenza, *A short proof of local regularity of distributional solutions of Poisson's equation*, Proc. Amer. Math. Soc., **148** (2020), no.5, 2143–2148.

[16] H. Dong, S. Kim, *On* C^1, C^2, *and weak type (1, 1) estimates for linear elliptic operators*, Comm. PDE, **42** (2017), 417–435.

[17] C. M. Elson, *An extension of Weyl's lemma to infinite dimensions*, Trans. Amer. Math. Soc., **194** (1974), 301–324.

[18] M. Erbar, K. Kuwada and K. Sturm, *On the equivalence of the entropic curvature-dimension condition and Bochner's inequality on metric measure spaces*, Invent. Math., **201** (2015), 993–1071.

[19] N. Gigli, *On the differential structure of metric measure spaces and applications*, Mem. Amer. Math. Soc., **236** (2015), no. 1113, 1–91.

[20] N. Gigli and E. Pasqualetto, *Lectures on Nonsmooth Differential Geometry*, Berlin: Springer, 2020.

[21] L. Hörmander, *Hypoelliptic second order differential operators*, Acta Math., **119** (1967), 147–171.

[22] B. Hua, M. Kell, and C. Xia, *Harmonic functions on metric measure spaces*, arXiv:1308.3607.

[23] R. Jiang, H. Li and H. Zhang, *Heat kernel bounds on metric measure spaces and some applications*, Potential Anal., **44** (2016), 601–627.

[24] T. Jin, V. Maz'ya and J. V. Schaftingen, *Pathological solutions to elliptic problems in divergence form with continuous coefficients*, C. R. Acad. Sci. Paris, Ser I, **347** (2009), 773–778.

[25] P. Li, *Uniqueness of L^1 solutions for the Laplace equation and the heat equation on Riemannian manifolds*, J. Differ. Geom., **20** (1984), 447–457.

[26] P. Li and R. Schoen, *L^p and mean value properties of subharmonic functions on Riemannian manifolds*, Acta Math., **153** (1984), 279–301.

[27] Y. Li, *On the C^1 regularity of solutions to divergence form elliptic systems with Dini-continuous coefficients*, Chin. Ann. Math., **38**B(2017), no.2 489–496.

[28] J. Lott and C. Villani, *Ricci curvature for metric-measure spaces via optimal transport*, Ann. Math., **169** (2009), no. 3, 903–991.

[29] V. G. Maz'ya, R. McOwen, *Differentiablilty of solutions to second-order elliptic equations via dynamical systems*, J. Diff. Eq., **250** (2011), 1137–1168.

[30] A. Mondino and A. Naber, *Structure theory of metric measure spaces with lower Ricci curvature bounds*, J. Eur. Math. Soc. (JEMS), **21** (2019), no. 6, 1809–1854.

[31] Y. Peng, X. Sun, H. C. Zhang, *A note on gradient estimates for elliptic equations with discontinuous coefficients*, Preprint.

[32] Y. Peng, H. C. Zhang, and X. P. Zhu, *Weyl's lemma on $RCD(K, N)$ metric measure spaces*. Preprint, arXiv:2212.09022.

[33] A. Petrunin, *Alexandrov meets Lott–Villani–Sturm*, Münst. J. Math. **4** (2011), 53–64.

[34] R. Schoen and S. T. Yau, *Lectures on Differential Geometry*, Boston: International Press, 1994.

[35] D. W. Strook, *Weyl's lemma, one of many//Groups and Analysis: The Legacy of Hermann Weyl*, New York: Cambridge University Press, 2008, 164–173.

[36] K. T. Sturm, *Analysis on local Dirichlet spaces. I. Recurrence, conservativeness and L^p-Liouville properties*. J. Reine Angew. Math., **456**(1994), 173–196.

[37] K. T. Sturm, *Analysis on local Dirichlet space-II. Upper Gaussian estimates for the fundamental solutions of parabolic equations*, Osaka J. Math., **32** (1995), 275–312.

[38] K. T. Sturm, *On the geometry of metric measure spaces. I*, Acta Math., **196** (2006), no. 1, 65–131.

[39] K. T. Sturm, *On the geometry of metric measure spaces. II*, Acta Math., **196** (2006), no. 1, 133–177.

[40] H. Weyl, *The method of orthogonal projection in potential theory*, Duke Math. J., **7** (1940), 411–444.

[41] H. C. Zhang and X. P. Zhu, *Ricci curvature on Alexandrov spaces and rigidity theorems*, Comm. Anal. Geom., **18** (2010), no. 3, 503–553.

[42] H. C. Zhang and X. P. Zhu, *Local Li-Yau's estimates on* $\mathsf{RCD}^*(K, N)$ *metric measure spaces*, Calc. Var. PDEs, **55** (2016), no. 4, Art. 93, 30 pp.

[43] W. Zhang and J. Bao, *Regularity of very weak solutions for elliptic equation of divergence form*, J. Funct. Anal., **262** (2012), 1867–1878.

Quadratic Hessian Equation

YU YUAN

Department of Mathematics, University of Washington, Seattle, WA, USA

E-mail: yuan@math.washington.edu

In memory of my teacher, Prof. Ding Wei-Yue (1945–2014)

1 Introduction

The elementary symmetric quadratic or σ_2 equation

$$\sigma_2\left(D^2u\right) = \sum_{1 \leqslant i_1 < i_2 \leqslant n} \lambda_{i_1}\lambda_{i_2} = \frac{\left(\triangle u\right)^2 - \left|D^2u\right|^2}{2} = 1 \tag{1.1}$$

with $\lambda_i's$ being the eigenvalues of the Hessian D^2u of scalar function u, is a nonlinear Hessian dependence equation of the lowest integer order, and is called fully nonlinear equation, because the nonlinearity is on the highest order derivatives of the solutions. The σ_2 equation sits in between the (linear) Laplace equation $\sigma_1\left(D^2u\right) = \lambda_1 + \cdots + \lambda_n = \triangle u = 1$ and the (fully nonlinear) Monge-Ampère equation $\sigma_n\left(D^2u\right) = \lambda_1\lambda_2\cdots\lambda_n = \det D^2u = 1$. The 2-sheet hyperboloid level set of the equation

$$\left\{ \lambda \in \mathbb{R}^n : \lambda_1 + \cdots + \lambda_n = \pm\sqrt{2 + |\lambda|^2} \right\}$$

is rotationally symmetric, unlike all the other σ_k equations with $3 \leqslant k \leqslant n$.

To make those equations elliptic, or monotone dependence on Hessian along positive definite symmetric matrices, we require the linearized operator positive definite. Equivalently, the normal of the level set has positive sign for all

This work is partially supported by NSF grant.

components in the eigenvalue space, respectively positive definite in the matrix space. For example, among all four branches of level set $\lambda_1\lambda_2\lambda_3 = 1$, only one is elliptic; the same is true for $\lambda_1\lambda_2\lambda_3 = -1$. The negative definite or all negative component case is also considered as elliptic. In particular, the two symmetric branches of $\sigma_2(\lambda) = 1$ are both elliptic. The choice of branch is made automatically by C^2 solutions. For less smooth solutions such as continuous viscosity ones, the positive $\triangle u > 0$ or negative $\triangle u < 0$ branch has to be specified. Afterward, $-D^2 u$ is on the other branch in the viscosity sense.

Replacing the flat eigenvalues $\lambda_i's$ with the principal curvatures $\kappa_i's$ of graph $(x, u(x))$ in Euclid space $\mathbb{R}^n \times \mathbb{R}^1$, one has the scalar curvature equation

$$\sigma_2(\kappa_1, \cdots, \kappa_n) = 1.$$

Recall $\kappa_i's$ are the eigenvalues of the normalized second fundamental form II by the induced metric g or shape matrix

$$IIg^{-1} = \frac{D^2 u}{\sqrt{1 + |Du|^2}}\left[I - \frac{Du \otimes Du}{1 + |Du|^2}\right] = \left[\partial_{x_i} A_{p_j}(Du)\right],$$

where $II = D^2 u/\sqrt{1 + |Du|^2}$, $g = I + Du \otimes Du$, and $A(p) = \sqrt{1 + |p|^2}$. Replacing the flat eigenvalues $\lambda_i's$ with the eigenvalues of Schouten tensor of a conformal metric $g = u^{-2}g_0$, one has σ_2-type Yamabe equation in conformal geometry, which simplifies to

$$\sigma_2\left(uD^2 u - \frac{1}{2}|Du|^2 I\right) = 1$$

for flat metric g_0. Replacing the flat eigenvalues $\lambda_i's$ with the eigenvalues of Hermitian Hessian $\partial\bar{\partial}u$, one has σ_2-type equation arising from complex geometry. Lastly, in three dimensions $\sigma_2(D^2 u) = 1$ or equivalently $\arctan\lambda_1 + \cdots + \arctan\lambda_3 = \pm\pi/2$ is the potential equation of minimal Lagrangian graph $(x, Du(x))$ with phase $\pm\pi/2$ in Euclid space $\mathbb{R}^3 \times \mathbb{R}^3$.

2　Results

2.1　Outline

Once an equation is given, the first question to answer is the existence of solutions. Smooth ones cannot be obtained immediately, in general; worse, they may not even exist. The typical approach is to first seek weak solutions, in the integral sense if the equation has divergence structure, or in the "pointwise integration by parts sense", namely, in the viscosity sense if the equation enjoys a comparison principle. After obtaining those weak solutions, one studies the regularity and other properties of the solutions, such as Liouville or Bernstein type rigidity for entire solutions. All these hinge on *a priori* estimates of derivatives of solutions:

$$\|D^2 u\|_{L^\infty(B_1)} \leqslant C(\|Du\|_{L^\infty(B_2)}) \leqslant C(\|u\|_{L^\infty(B_3)}).$$

Having the L^∞ bound of the Hessian available, the ellipticity of the above fully nonlinear equations becomes uniform, we can apply the Evans-Krylov-Safonov theory (for the ones with convexity/concavity, possibly without divergence structure) or the Evans-Krylov-De Giorgi-Nash theory (for the ones with convexity/concavity and divergence structure) to obtain $C^{2,\alpha}$ estimates of solutions. Either theory can handle the quadratic Hessian equation along with all other σ_k equations, because they all share the divergence structure. In the σ_2 equation (1.1) case, the linearized operator is readily seen in divergence form

$$\triangle_F = \sum_{i,j=1}^n F_{ij}\partial_{ij} = \sum_{i,j=1}^n \partial_i \left(F_{ij}\partial_j\right),$$

with

$$(F_{ij}) = \triangle u\, I - D^2 u = \sqrt{2 + |D^2 u|^2}\, I - D^2 u > 0.$$

Here and in the remaining, for certainty, and without loss of generality, we assume $\triangle u > 0$. The concavity of the equation is evident in an equivalent form

$$\triangle u - \sqrt{2 + |D^2 u|^2} = 0, \tag{2.1}$$

whose linearized operator

$$I - \frac{D^2 u}{\sqrt{2 + |D^2 u|^2}}$$

also reveals the uniform ellipticity of the equation with uniformly bounded Hessian solutions.

Considering the minimal surface structure of the quadratic Hessian equation in three dimensions, the $C^{2,\alpha}$ estimate can also be achieved via geometric measure theory. For the σ_n or Monge-Ampère equation, back in the 1950s, Calabi attained C^3 estimates by interpreting the cubic derivatives in terms of the curvature of the corresponding Hessian metric $g = D^2 u$. Further, iterating the classic Schauder estimates, one obtains smoothness of the solutions, and even analyticity, if the smooth equations such as all the σ_k equations are also analytic.

2.2 Rigidity of entire solutions

The classic Liouville type theorem asserts every entire harmonic function bounded from below or above is a constant by the Harnack inequality. Thus every semiconvex harmonic function is a quadratic one, as its double derivatives are all harmonic with lower bounds, hence constants. Similarly, every entire (convex) solution to the Monge-Ampère equation $\det D^2 u = 1$ is quadratic. This was first proved in two-dimensional case by Jörgens, later in dimension up to five by Calabi, and in all dimensions by Pogorelov. Also, Cheng-Yau had a geometric proof.

Recently, Shankar-Yuan [SY3] proved that every entire semiconvex solution to the quadratic Hessian equation $\sigma_2\left(D^2 u\right) = 1$ is quadratic. In dimension two, it is the above classic Jörgens's theorem without any extra condition (not even convexity) on the entire solutions, thus a Bernstein type result. In three dimensions, this was proved in [Y] earlier, as a by-product of rigidity for the special Lagrangian equation.

Under an almost convexity condition on entire solutions to $\sigma_2\left(D^2 u\right) = 1$ in general dimensions, Chang-Yuan derived the rigidity [ChY]. Under a general semiconvexity and an additional quadratic growth assumption on entire solutions in general dimensions, Shankar-Yuan showed the rigidity in [SY1]. Assuming only quadratic growth on entire solutions to $\sigma_2\left(D^2 u\right) = 1$ in three and four dimensions, the same rigidity result was proved in the joint work with Warren [WY] and Shankar [SY4] respectively. Assuming a super quadratic growth condition, Bao-Chen-Ji-Guan [BCGJ] demonstrated that all convex entire solutions to $\sigma_2\left(D^2 u\right) = 1$ along with other $\sigma_k\left(D^2 u\right) = 1$ are quadratic polynomials; and Chen-Xiang [CX][showed that all "super quadratic" entire solutions to $\sigma_2\left(D^2 u\right) = 1$ with $\sigma_1\left(D^2 u\right) > 0$ and $\sigma_3\left(D^2 u\right) \geqslant -K$ are also quadratic polynomials.

Warren's rare saddle entire solution $u(x_1, \cdots, x_n) = (x_1^2 + x_2^2 - 1) e^{x_3} + \frac{1}{4} e^{-x_3}$ to $\sigma_2(D^2 u) = 1$ in dimension three and above [W], confirms the necessity of the semiconvexity or the quadratic growth assumption. C.-Y. Li [L] followed with "non-degenerate" entire solution $u(x) = (x_1^2 + x_2^2 - 1) e^{x_n} + \frac{n-2}{4} e^{-x_n} + (x_3 + \cdots + x_{n-1}) x_n$ in dimension n and above for $n \geqslant 4$.

2.2.1 Two dimensions

In the following, we recap Nitsche's idea in showing the rigidity of entire solutions in two dimensions.

Given a C^2 solution u to $\sigma_2(D^2 u) = 1$, up to negation, we assume $D^2 u$ is on the positive branch of the hyperbola $\lambda_1 \lambda_2 = 1$, in turn, u is convex. Let w be the Legendre-Lewy transform of $u(x)$, that is, the Legendre transform of $u(x) + |x|^2/2$. Geometrically their "gradient" graphs satisfy $(x, Du(x) + x) = (Dw(y), y) \in \mathbb{R}^2 \times \mathbb{R}^2$, and the "slopes" of graphs satisfy

$$(I, D^2 u(x) + I) = \left(D^2 w(y) \frac{\partial y}{\partial x}, \frac{\partial y}{\partial x} \right).$$

It follows that

$$I < D^2 u(x) + I = \left(D^2 w(y) \right)^{-1} \quad \text{or} \quad D^2 u(x) = \left(D^2 w(y) \right)^{-1} - I.$$

Taking determinants yields

$$1 = \sigma_2(D^2 u) = \det \left[\left(D^2 w(y) \right)^{-1} - I \right] = \mu_1^{-1} \mu_2^{-1} - \mu_1^{-1} - \mu_2^{-1} + 1.$$

or an equation for the eigenvalues $\mu_i's$ of $D^2 w$

$$1 = \mu_1 + \mu_2 = \Delta w.$$

Noticing the boundedness of Hessian $D^2 w$

$$0 < D^2 w < I,$$

we see the constancy by Liouville. Consequently from the flatness of graphs $(x, Du(x) + x) = (Dw(y), y)$ or constancy of $(D^2 w)^{-1} - I = D^2 u$, it follows that u is quadratic. Note that Jörgens' original "involved" proof made use of a partial Legendre transformation.

Going further, this Legendre-Lewy transformation proof of Jörgens' theorem, coupled with Heinz transformation, led Nitsche [N] to his elementary proof of the original Bernstein theorem: every entire solution to the minimal surface equation

$$\text{div} \left(\frac{Df}{\sqrt{1 + |Df|^2}} \right) = 0$$

in two dimensions is linear.

The remaining proof goes as follows. The mean curvature vector $\vec{H}$ of the graph $(x_1, x_2, f(x))$ is

$$\vec{H} = \triangle_g (x_1, x_2, f(x)) = \sum_{i,j} \frac{1}{\sqrt{g}} \partial_i \left(\sqrt{g} g^{ij} \partial_j \right) (x_1, x_2, f(x))$$

$$= \left(\text{div} \frac{(1 + f_2^2, -f_1 f_2)}{\sqrt{1 + |Df|^2}}, \text{div} \frac{(-f_1 f_2, 1 + f_1^2)}{\sqrt{1 + |Df|^2}}, \text{div} \frac{(f_1, f_2)}{\sqrt{1 + |Df|^2}} \right),$$

where $\sqrt{g} g^{-1} = \begin{bmatrix} 1 + f_2^2 & -f_1 f_2 \\ -f_2 f_1 & 1 + f_1^2 \end{bmatrix} / \sqrt{1 + |Df|^2}$ was used in the last equality.

The magnitude H of the mean curvature vector is, as noted before

$$H = Tr \left[\partial_{x_i} A_{p_j} (Df) \right] = \text{div} \frac{(f_1, f_2)}{\sqrt{1 + |Df|^2}} = 0.$$

Hence, $\triangle_g (x_1, x_2, f(x)) = (0, 0, 0)$.

Considering the first component equation $\triangle_g x_1 = 0$, the conjugate function of x_1 is defined as

$$x_1^* (x_1, x_2) = \int^{(x_1, x_2)} \frac{f_1 f_2 dx_1 + (1 + f_2^2) dx_2}{\sqrt{1 + |Df|^2}}.$$

Similarly, the conjugate function of x_2 is also defined. Together, they represent the "normalized" metric

$$\frac{1}{\sqrt{1 + |Df|^2}} \begin{bmatrix} 1 + f_1^2 & f_1 f_2 \\ f_2 f_1 & 1 + f_2^2 \end{bmatrix} = \begin{bmatrix} Dx_2^* \\ Dx_1^* \end{bmatrix}.$$

By symmetry of the left-side matrix, $\partial_2 x_2^* = \partial_1 x_1^*$, then there exists a double potential u so that $Du = (x_2^*, x_1^*)$. Thus, one has Heinz transformation u of the height function f of a minimal graph satisfying

$$\frac{1}{\sqrt{1+|Df|^2}} \begin{bmatrix} 1+f_1^2 & f_1 f_2 \\ f_2 f_1 & 1+f_2^2 \end{bmatrix} = D^2 u \quad \text{and} \quad \det D^2 u = 1 \quad \text{on } \mathbb{R}^2.$$

As just obtained, the Hessian $D^2 u$ of entire solution u is a constant matrix, and in turn, Df is a constant vector. The original Bernstein theorem is reached.

In passing, let us note the conjugate function of f

$$f^*(x_1, x_2) = \int^{(x_1, x_2)} \frac{-f_2 dx_1 + f_1 dx_2}{\sqrt{1+|Df|^2}}$$

satisfies

$$Df^* = \frac{(-f_2, f_1)}{\sqrt{1+|Df|^2}} \quad \text{and} \quad \sqrt{1-|Df^*|^2} = \frac{1}{\sqrt{1+|Df|^2}} \in (0,1)$$

or

$$\frac{Df^*}{\sqrt{1-|Df^*|^2}} = (-f_2, f_1).$$

Consequently

$$\operatorname{div}\left(\frac{Df^*}{\sqrt{1-|Df^*|^2}}\right) = 0.$$

Observe that the above conjugation process from minimal surface to maximal surface is reversible. "Incidentally" we have obtained a two-dimensional Bernstein type result: every entire solution to the maximal surface equation $\operatorname{div}\left(Df/\sqrt{1-|Df|^2}\right) = 0$ is linear, which was first proved up to four dimensions by Calabi, and in general dimensions by Cheng-Yau.

2.2.2 General dimensions

Next, we outline the argument toward rigidity for semiconvex entire solutions to $\sigma_2(D^2 u) = 1$ in general dimensions.

The Legendre-Lewy transform of a general semiconvex solution satisfies a uniformly elliptic, saddle equation with bounded Hessian. In the almost convex case, the new equation becomes concave, thus the Evans-Krylov-Safonov theory yields the constancy of the bounded new Hessian, and in turn, the old one. To beat the saddle case, one has to be "lucky", only one time. Recall that, in general the Evans-Krylov-Safonov fails as shown by the saddle counterexamples of Nadirashvili-Vladuts. Our earlier trace Jacobi inequality, as an alternative log-convex vehicle, other than the maximum eigenvalue Jacobi inequality, in deriving the Hessian estimates for general semiconvex solutions [SY1], could rescue the saddleness. But the trace Jacobi only holds for large enough trace of the Hessian. It turns out that the trace added by a large enough constant satisfies the elusive Jacobi inequality.

Equivalently, the reciprocal of the shifted trace Jacobi quantity is superharmonic, and it remains so in the new vertical coordinates under the Legendre-Lewy transformation by a transformation rule. Then the iteration arguments developed in our joint work with Caffarelli show the "vertical" solution is close to a "harmonic" quadratic at one small scale, "luckily" (two steps in the execution: the superharmonic quantity concentrates to a constant in measure by applying Krylov-Safonov's weak Harnack; a variant of the superharmonic quantity, as a quotient of symmetric Hessian functions of the new potential, is very pleasantly concave and uniformly elliptic, consequently, closeness to a "harmonic" quadratic is possible by the Evans-Krylov-Safonov theory), and the closeness improves increasingly as we rescale (this is a self-improving feature of elliptic equations, no concavity/convexity needed). Thus a Hölder estimate for the bounded Hessian is realized, and consequently so is the constancy of the new and then the old Hessian.

Note that, in three dimensions, our proof provides a "pure" PDE way to establish the rigidity, distinct from the geometric measure theory way used in our earlier work on the rigidity for special Lagrangian equations two decades ago.

The details are in the following.

Step 1.   Bounded Hessian and uniform ellipticity after Legendre-Lewy transform.

The Legendre-Lewy transform $w(y) = \mathcal{L}\mathcal{L}\left[u(x) + K|x|^2/2\right]$ of a general semiconvex solution $u(x)$ with $D^2u \geqslant (\delta - K)I$ satisfies a uniformly elliptic, saddle equation with bounded Hessian:

$$(x, Du(x) + Kx) = (Dw, y),$$

$$0 < D^2w = (D^2u + K)^{-1} < \delta^{-1} \text{ or } \lambda_i = \mu_i^{-1} - K \geqslant \delta - K,$$

$$g(\mu) = -f(\mu^{-1} - K) = -\sigma_2(\mu^{-1} - K) = -1.$$

By Lin-Trudinger [LT], and also Chang-Yuan [ChY]

$$\lambda_1^{-1} \lesssim f_{\lambda_1} \lesssim \lambda_1, \quad f_{\lambda_{k \geqslant 2}} \approx \lambda_1 \text{ for } \lambda_1 \geqslant \cdots \geqslant \lambda_n;$$

for $\sigma_2(\lambda) = 1$ with $\lambda_i \geqslant \delta - K$, all but one eigenvalues are bounded, $|\lambda_{k \geqslant 2}| \leqslant C(K)$ and $f_{\lambda_1}\lambda_1 \approx 1$, then

$$g_{\mu_i} = f_{\lambda_i}\mu_i^{-2} = f_{\lambda_i}(\lambda_i + K)^2 \approx C(n,K)\,\lambda_1.$$

Consequently, level set

$$\{\mu \mid g(\mu) = -\sigma_2(\mu^{-1} - K) = -1\} \text{ is a uniformly elliptic surface.} \qquad (2.2)$$

The new equation also takes the form

$$\frac{\sigma_{n-2}(\mu)}{\sigma_n(\mu)} - (n-1)K\frac{\sigma_{n-1}(\mu)}{\sigma_n(\mu)} + \frac{n(n-1)}{2}K^2 = 1.$$

Remark In the almost convex case $K = \sqrt{2/n(n-1)}$, the new equation becomes

$$\frac{\sigma_{n-1}(\mu)}{\sigma_{n-2}(\mu)} = [(n-1)K]^{-1}, \qquad (2.3)$$

thus concave. Then the Evans-Krylov-Safonov theory yields the constancy of the bounded new Hessian

$$[D^2w]_{C^\alpha(B_R)} \leqslant \frac{C(n)}{R^\alpha}\|D^2w\|_{L^\infty(B_{2R})} \leqslant \frac{C(n)}{R^\alpha} \to 0 \text{ as } R \to \infty,$$

and in turn, the old one.

The uniformly elliptic level set $\{g(\mu) = -\sigma_2(\mu^{-1} - K) = -1\}$ is saddle for large K. There are $C^{1,\alpha}$ singular solutions to uniformly elliptic saddle equation in five dimensions by Nadirashvili-Tkachev-Vladuts.

In the next two steps, we really use the following equivalent uniformly elliptic saddle equation

$$H\left(D^2 w\right) = \sigma_n\left(\mu\right)\left[1 - \sigma_2\left(\mu^{-1} - K\right)\right] = 0,$$

which in three dimensions, becomes

$$-\sigma_1\left(\mu\right) + 2K\sigma_2\left(\mu\right) - \left(3K^2 - 1\right)\sigma_3\left(\mu\right) = 0 \quad \text{or}$$

$$\frac{\sigma_2\left(\mu\right)}{\sigma_1\left(\mu\right)} = \left[2K - \left(3K^2 - 1\right)\frac{\sigma_3\left(\mu\right)}{\sigma_2\left(\mu\right)}\right]^{-1}.$$

Step 2. Shifted trace Jacobi inequality to rescue saddleness.

For $F\left(D^2 u\right) = \sigma_2\left(\lambda\right) = 1$ with $D^2 u \geqslant -KI$, $b\left(x\right) = \ln\left(\triangle u + nK\right)$ satisfies the elusive strong subharmonicity

$$\triangle_F b = F_{ij}\partial_{ij} b \geqslant F_{ij}\partial_i b\,\partial_j b = |\nabla_F b|^2$$

which is equivalent to the superharmonicity, by a transformation rule in [SY1].

$$\triangle_H a\left(y\right) \leqslant 0$$

$$\text{with } a\left(y\right) = \frac{1}{\lambda_1 + K + \cdots + \lambda_n + K} = \frac{\sigma_n\left(\mu\right)}{\sigma_{n-1}\left(\mu\right)} \overset{n=3}{=} \frac{\sigma_3\left(\mu\right)}{\sigma_2\left(\mu\right)}.$$

Remark For log-convex function $b = \ln\lambda_{\max}$ or $\ln\left(\lambda_{\max} + K\right)$, Jacobi inequality $\triangle_F b \geqslant |\nabla_F b|^2$ holds in three dimensions without any restriction, in general dimensions with necessary semiconvexity condition. The reciprocal $\mu_{\min}\left(y\right) = \left(\lambda_{\max} + K\right)^{-1}$ is superharmonic $\triangle_H \mu_{\min}\left(y\right) \leqslant 0$. But $\mu_{\min}\left(D^2 w\right)$ is not a uniformly elliptic function/operator on $D^2 w$, though concave. Therefore, it is not adequate via $\mu_{\min}\left(y\right)$ to run the Caffarelli-Yuan procedure for Hölder of Hessian $D^2 w$.

For log-linear $b = \ln\triangle u$, Qiu showed $\triangle_F b \overset{3\text{-d}}{\geqslant} |\nabla_F b|^2$ in [Q]. It is indeed another log-convex function

$$\ln\triangle u = \ln\sqrt{|\lambda|^2 + 2} \mod \sigma_2\left(\lambda\right) = 1$$

satisfying $\triangle_F \ln\triangle u \geqslant |\nabla_F \ln\triangle u|^2$ for large enough $\triangle u \geqslant C\left(K\right)$ under semiconvexity $D^2 u \geqslant -K$. It is important to have the above shifted Jacobi quantity $b(x) = \ln(\triangle u + nK)$ valid without assuming $\triangle u$ large enough, to execute our argument toward constancy of the Hessian.

Remark The above Jacobi inequality is actually an equality on minimal surface $(x, f(x)) \in \mathbb{R}^n \times \mathbb{R}^1$ for $\mathrm{div}\left(Df/\sqrt{1 + |Df|^2}\right) = 0$:

$$\triangle_g b = |\nabla_g b|^2 + \left|IIg^{-1}\right|^2 \quad \text{or} \quad \triangle_g \omega = -\omega \left|IIg^{-1}\right|^2 \leqslant 0,$$

where $b = \ln\sqrt{1 + |Df|^2}$, and $\omega = \langle (0, \cdots, 0, 1), N \rangle = 1/\sqrt{1 + |Df|^2}$ is the effective deformation, while varying the minimal surface along a Jacobi vector field $J = (0, \cdots, 0, 1)$.

Step 3. Hölder estimate of new Hessian on saddle equation.

We illustrate the argument for Hölder estimates for Hessian on uniformly elliptic saddle equation in three dimensions, where the idea is not lost, but the gain is a better understanding of the idea.

Again, the new "vertical" uniformly elliptic saddle equation is

$$\frac{\sigma_2(\mu)}{\sigma_1(\mu)} = \left[2K - (3K^2 - 1)\frac{\sigma_3(\mu)}{\sigma_2(\mu)}\right]^{-1}$$

for $1 \geqslant \mu_{i \geqslant 2} \geqslant c(K) > 0$ and $1 \geqslant \mu_1 > 0$.

Apply Krylov-Safonov's weak Harnack to the bounded superharmonic $a(y) = \dfrac{\sigma_3(\mu)}{\sigma_2(\mu)}$ from Step 2, $a(y)$ concentrates to a level $l = \min\limits_{B_\varepsilon} a(y)$ in one small (enough) ball $B_\varepsilon(0)$, that is $\dfrac{\sigma_3(\mu)}{\sigma_2(\mu)} \approx l$ in 99.99999% of B_ε. Note the concave $\dfrac{\sigma_3(\mu)}{\sigma_2(\mu)}$ is not uniformly elliptic, because μ_1 could be close 0.

Step 3 Continued: Remarkably, μ is approximately on the uniformly elliptic $(1 \geqslant \mu_3, \mu_2 \geqslant c(K) > 0)$, concave (level set $\dfrac{\sigma_2(\mu)}{\sigma_1(\mu)} = l$ is a concave surface) equation

$$\frac{\sigma_2(\mu)}{\sigma_1(\mu)} = \left[2K - (3K^2 - 1)\, l\,\right]^{-1} \quad \text{in } 99.99999\% \text{ of } B_\varepsilon$$

By the existence of Dirichlet problem via Evans-Krylov-Safonov and Alexandrov maximum principle in measure,

$$w(y) = \text{quadratic } Q(y) \pm 0.00000000001 \text{ in } B_{\varepsilon/2}.$$

By self-improving property of the smooth uniformly elliptic equation (no concavity needed)

$$-\sigma_1(\mu) + 2K\sigma_2(\mu) - (3K^2 - 1)\sigma_3(\mu) = 0,$$

through iterative procedure $w(y) \approx y^{2+\alpha}$ near $y = 0$. Similarly near everywhere in $B_{\varepsilon/4}$. Thus Hölder estimates for D^2w.

Finally, by quadratic scaling

$$[D^2w]_{C^\alpha(B_R)} \leqslant \frac{C(n, K)}{R^\alpha} \overset{R \to \infty}{\longrightarrow} 0.$$

Thus D^2w is constant matrix, in turn, so is D^2u.

Remark　In the original Caffarelli-Yuan procedure, the superharmonic quantity $a = \dfrac{\sigma_3(\mu)}{\sigma_2(\mu)}$ is $1 - e^{K\Delta u}$ for saddle uniformly elliptic equation $F(D^2u) = 0$ with convex level sets along trace trM of the level set $\{F(M) = 0\}$. Once, only one lucky chance needed, Δu concentrates in measure, the solution u is close to a quadratic, by solving the Laplace equation and applying the Alexandrov maximum principle in measure. Afterwards, the self-improving machinery of smooth uniformly elliptic equation takes over for Hölder estimate of Hessian.

2.3　Regularity for viscosity solutions

Any reasonable solution to the σ_1 or Laplace equation $\Delta u = 1$ such as continuous viscosity in pointwise sense or distributional solution in the integral sense is analytic. The same is not true for $\sigma_{k \geqslant 3}$ equations, and unknown for σ_2 equation in dimension five and higher currently. In fact, by now there are $C^{1,\varepsilon}$ and Lipschitz Pogorelov-like singular viscosity solutions to $\sigma_{k \geqslant 3}$ equations in dimension three and higher. In the σ_n or Monge-Ampère equation case, those viscosity solutions are also singular solutions in the Alexandrov integral sense.

The advance in the joint work with Chen-Shankar [CSY] also led us to obtain interior regularity (analyticity) for almost convex viscosity solutions to the quadratic Hessian equation (1.1), in the joint work with Shankar [SY2]. Due to similar conceptual and technical challenges—smooth approximations may not preserve those semiconvexity constraints—we cannot invoke our available Hessian estimates with Shankar [SY1] for general semiconvex solutions or with McGonagle and Song [MSY] for almost convex solutions, while taking the limit and deduce interior regularity. A key observation is that the Legendre-Lewy transform of any semiconvex viscosity solution to the equivalent concave equation (2.1) stays viscosity solution to a new concave (2.3) (only for the original almost convex solution) and uniformly elliptic equation (2.2) (for all original semiconvex solutions). In passing, let us note that, for a general fully nonlinear second order elliptic equation, Alvarez-Lasry-Lions showed that the Legendre transform of any strictly convex C^2 solution is a convex viscosity solution of a conjugate equation. Moreover, the "striking" role of the C^2 regularity of the original solution in their arguments was pointed out [ALL, p.281]. It follows that the transformed $C^{1,1}$ solution is smooth by the Evans-Krylov-Safonov theory. Then the boundedness of the original solutions combined with the constant rank theorem by Caffarelli-Guan-Ma [CGM] implies that the original viscosity solution is smooth.

Shortly after, Mooney [M] provided a different proof of the interior regularity for convex viscosity solutions: every such convex solution is strictly 2-convex, then all smooth approximated solutions enjoy uniform Pogorelov-type $C^{1,1}$ and higher derivative estimates by Chou-Wang [CW], in turn, the interior regularity by taking limit.

In two dimensions, the above regularity result (now the convexity condition is automatic) actually also follows from Heinz's famous Hessian estimate earlier. In three dimensions, the regularity for continuous viscosity solutions to (1.1) follows from the Hessian estimate in our joint work with Warren [WY] and smooth existence with smooth boundary value by Caffarelli-Nirenberg-Spruck, and also Trudinger. Our most recent joint work with Shankar [SY4] on Hessian estimates for the quadratic Hessian equation (1.1) in four dimensions yields up the same regularity in four dimensions. There, a direct way to interior regularity without first deriving the Hessian estimates is also provided. Consequently, a compactness argument leads to an implicit Hessian estimate.

2.4　A priori Hessian estimates

In our long investigation, culminating in the most recent joint work with Shankar [SY4], we obtained an implicit Hessian estimate and interior regularity (analyticity) for the quadratic Hessian equation $\sigma_2 \left(D^2 u \right) = 1$ in four dimensions. Our compactness method (almost Jacobi inequality–doubling–twice differentiability–small perturbation) also provides respectively a Hessian estimate for smooth solutions satisfying a dynamic semiconvexity condition in higher dimensions, which includes convexity, almost convexity, and semiconvexity conditions appeared in the recent papers on Hessian estimates, and a non-minimal surface proof for the corresponding three dimensional results in our earlier joint work with Warren [WY].

Other consequence is a rigidity result for entire solutions to the σ_2 equation with quadratic growth, namely all such solutions must be quadratic, provided the smooth solutions in dimension $n \geqslant 5$ also satisfying the dynamic semiconvex assumption.

Again, the Hessian estimate for the σ_2 or Monge-Ampère equation in dimension two was achieved by Heinz in the 1950s. Hessian estimates fail for the Monge-Ampère equation in dimension three and higher, as illustrated by the famous counterexamples of Pogorelov in the 1970s; those irregular solutions also serve as counterexamples for cubic and higher order symmetric $\sigma_{k \geqslant 3}$ equations, for example by Urbas. Hessian estimates for solutions with certain strict convexity constraints to the Monge-Ampère and $\sigma_{k \geqslant 2}$ equations were derived by Pogorelov and later Chou-Wang respectively using the Pogorelov technique; some (pointwise) Hessian estimates in terms of certain integrals of the Hessian were obtained by Urbas in the early 2000s. The gradient estimates for σ_k equations were derived by Trudinger, Chou-Wang in the mid 1990s.

The compactness proof toward an implicit Hessian estimate for almost convex solutions in [MSY] is based on the concavity of uniformly elliptic equation (2.3) under the Legendre-Lewy transformation, a constant rank theorem by Caffarelli-Guan-Ma [CGM], on the vertical side, and a strip argument on the horizontal side.

The proof toward an explicit Hessian estimate for semiconvex solutions is based on an elusive-Jacobi inequality-satisfying quantity, the maximum eigenvalue of the Hessian of the solutions, envisioned to be true in 2012. Another essential new device is a mean value inequality for the strongly subharmonic maximum eigenvalue under the Legendre-Lewy transformation with uniformly elliptic equation (2.2), and its weighted version converted back to the original variables or horizontal side.

The new idea for Hessian estimates, under a dynamic semiconvexity condition in dimension five and higher, and consequently interior regularity in dimension four, is first to get a doubling, or a "three-sphere" inequality for the Hessian bound on the middle ball, in terms of Hessian bound on a small inner ball and gradient bound on the outer large ball:

$$\max_{B_2(0)} \triangle u \leqslant C\left(r, \|u\|_{Lip(B_3(0))}\right) \max_{B_r(0)} \triangle u.$$

Using a Jacobi inequality, true with a lower σ_3-bound condition for Hessian, satisfied by convex solutions, Guan-Qiu [GQ] reached their Hessian estimate for the quadratic Hessian equation. Qiu [Q] followed with his doubling in three dimensions, where the Jacobi inequality was unconditionally available since [WY]. But the maximum of Guan-Qiu test function

$$P = 2\ln\left(9 - |x|^2\right) + \alpha |Du|^2/2 + \beta\ (x \cdot Du - u)$$

$$+ \ln\max\left\{\ln\frac{\triangle u}{\max_{B_1(0)} \triangle u}, \gamma^{-1}\right\}$$

with small $\gamma = \gamma(n) > 0$, smaller $\beta = \beta\left(\gamma, \|u\|_{Lip(B_3(0))}\right) > 0$

and smallest $\alpha = \alpha\left(\gamma, \|u\|_{Lip(B_3(0))}\right) > 0,$

could not be ruled out from happening on the small inner ball without the σ_3-lower bound assumption, thus Qiu's "three-sphere" inequality.

Now only an almost Jacobi inequality is available in dimension four. In fact, as observed in 2012, there is no Jacobi inequality in dimension four, and worse, even no subharmonicity of the Laplace of the log of Hessian in dimension five, thus the added dynamic semiconvexity condition in higher dimensions for an almost Jacobi inequality:

$$\triangle_F b \overset{n=4}{\geqslant} \left(\frac{1}{2} + \frac{\lambda_{\min}}{\triangle u}\right) |\nabla_F b|^2 \geqslant 0;$$

$$\triangle_F b \overset{n\geqslant 5}{\geqslant} \left(c_n + \frac{\lambda_{\min}}{\triangle u}\right) |\nabla_F b|^2 \geqslant 0, \quad \text{if} \quad c_n + \frac{\lambda_{\min}}{\triangle u} \geqslant 0$$

$$\text{with} \quad c_n = \frac{\sqrt{3n^2+1}-n+1}{2n} \quad \text{and} \quad b = \ln \triangle u$$

Note that for $\sigma_2(\lambda) = 1$, we have $\dfrac{\lambda_{\min}}{\triangle u} > -\dfrac{n-2}{n}$, and at the extreme configuration $\lambda = \left(K, \cdots, K, -\dfrac{n-2}{2}K + \dfrac{1}{(n-1)K}\right)$, one has $\dfrac{\lambda_{\min}}{\triangle u} > -\dfrac{n-2}{n}$. In fact Jacobi inequality holds $\triangle_F b \overset{n=3}{\geqslant} \dfrac{1}{3}|\nabla_F b|^2 \geqslant \left(\dfrac{1}{2} - \dfrac{1}{3}\right)|\nabla_F b|^2$ unconditionally in three dimensions.

But the almost Jacobi inequality is really a regular one away from the extreme configuration of the equation, where the equation is conformally uniformly elliptic. Qiu's doubling argument can be pushed through.

Now to find a small inner ball where the Hessian is bounded, we first show the almost everywhere twice differentiability of continuous viscosity solutions,

$$u(x) - Q_y(x) = o\left(|x-y|^2\right)$$

by adapting Chauder-Trudinger's argument [CT] for k-convex functions with $k > n/2$, with the gradient estimates $\|Du\|_{L^\infty(B_1)} \leqslant C(n) \|u\|_{L^\infty(B_2)}$, actually its integral form of control (a Hölder substitute for k-convex function was used in [CT]) for σ_k equations in [T] and also [CW], and the fact that $D^2 u$ is a bounded Borel measure for solutions of $\sigma_2(D^2 u) = 1$, as

$$\int_{B_1} |D^2 u|\, dx \overset{\triangle u = \sqrt{2+|D^2 u|^2}}{<} \int_{B_1} \triangle u\, dx \leqslant C(n)\|Du\|_{L^\infty(B_1)}.$$

Then Savin's small perturbation (from the quadratic polynomial at a twice differentiable point) [S] guarantees the small inner ball with bounded Hessian.

In our most recent follow-up paper with Shankar [SY5], a new proof of regularity for strictly convex solutions to $\det D^2 u = 1$ is found, using similar doubling methods, instead of Euclid distance, now in terms of an extrinsic distance on the maximal Lagrangian submanifold determined by the potential Monge-Ampère equation. This "strict convex" regularity was achieved originally by Pogorelov in the 1960s and 1970s, and generalized by Urbas and Caffarelli in the late 1980s.

3 Problems

Problem 1 Are there singular (Lipschitz) viscosity solutions, $W^{2,1}$ regularity, and any better partial regularity $\sigma_2 \left(D^2 u \right) = 1$ in dimension five or higher?

Given the Jacobi inequality is exhausted in our argument for Hessian estimates and regularity in four dimensions, it is time to look for singular viscosity solutions and better partial regularity for possible singular viscosity solutions in dimension five or higher. For example, a dimension estimate on the singular set of possible singular viscosity solutions. Note that by the gradient estimate, and then smooth approximations in Lipschitz norm, all continuous viscosity solutions are Lipschtiz, and by our almost everywhere twice differentiability [SY4] and Savin's small perturbation [S], the possible singular set is closed and with zero Lebesgue measure (Also true for viscosity solutions to $\sigma_{k \geqslant 3}$ equation in dim $n \geqslant 3$).

From $\displaystyle\int_{B_l} \left| D^2 u \right| dx \overset{\sigma_2 = 1}{<} \int_{B_l} \triangle u \, dx \leqslant \|Du\|_{L^\infty(\partial B_l)} |\partial B_l|$ and the gradient estimate, $D^2 u$ is a bounded Borel measure. It is reasonable to expect a $W^{2,1}$ regularity in dim $n \geqslant 5$. Recall that, unlike function $|x_1|$, all (convex) viscosity solutions to $\sigma_n \left(D^2 u \right) = 1$ have been shown to be $W^{2,1}$.

Problem 2 Regularity for semiconvex viscosity solutions to $\sigma_2 \left(D^2 u \right) = 1$ in dimension five or higher.

It is still unclear to us whether semiconvex viscosity solutions are regular, if only $D^2 u \geqslant -KI$ for large $K > 0$. The Legendre-Lewy transform is still a $C^{1,1}$ viscosity solution of a new uniformly elliptic equation (2.2), for any semiconvex viscosity solution. However, as the new equation no longer has convex level set, for large K, we are unable to deduce smoothness for the transformed solution at this point. Without the smoothness, we are currently unable to obtain a $C^{1,1}$ version of the constant rank theorem to gain positive definiteness of the semi-positive Hessian, for the $C^{1,1}$ solution of a uniformly elliptic and inversely convex equation on the vertical side. Otherwise, the interior regularity for such semiconvex viscosity solutions would be justified. At this point, it appears a far stretch to reach regularity for dynanic semiconvex viscosity solutions in dimension five or higher.

One follow-up of our Hessian estimates for three and the very recent four dimension σ_2 equation would be

Problem 3 Derive Schauder and Calderón-Zygmund estimates for variable-right-hand-side equation $\sigma_2 \left(D^2 u \right) = f \left(x \right)$ in dimension four.

With a $C^{1,1}$ assumption on $f \left(x \right)$, Qiu [Q] has generalized the arguments in [WY] to obtain Hessian estimates in dimension three. With an almost sharp Lipschitz assumption on $f \left(x \right)$, very recently, Zhou [Z] reached the Hessian estimate in three dimensions, along his Hessian estimates for the "twist" special Lagrangian equation $\sum\limits_{i=1}^{n} \arctan \lambda_i / f \left(x \right) = c \in \left[(n-2) \pi/2, n\pi/2 \right)$.

Consequently $C^{2,\alpha}$ estimates follow. Under a small enough Hölder seminorm assumption on f, Xu [X] derived interior $C^{2,\alpha}$ estimates in dimension three. Notice that the interior gradient estimates in [T] and [CW] needs Lipschitz assumption on f. The subtle small seminorm constraint is due to the non-uniform elliptic nature of the equation. The method works in general dimensions such as the recent four dimensions, as long as the interior Hessian estimate is available for the quadratic Hessian equation with constant right-hand side.

Problem 4 Any "elementary" pointwise argument toward Hessian estimates for $\sigma_2 \left(D^2 u \right) = 1$ in dimension three and four, in general higher dimension with the dynamic semiconvexity condition, as in the two dimensional case by Chen-Han-Ou [CHO], and convex case by Guan-Qiu [GQ]?

Furthermore, any explicit Hessian bound in terms of the gradient, as the quadratic exponential dependence in dimension three and general semiconvex case?

To gain more understanding of σ_2 equation, one distinct double divergence structure of σ_2 from $\sigma_{k \geqslant 3}$ is worth studying.

Problem 5 Under what additional condition on $u \in W^{1,2} \left(\Omega \right)$ does the equation $\sigma_2 \left(D^2 u \right) = 1$ in the very weak sense ([I]):

$$\int_\Omega \sum_{1 \leqslant i < j \leqslant n} \left(\varphi_{ij} u_i u_j - \frac{1}{2} \varphi_{ii} u_j^2 - \frac{1}{2} \varphi_{jj} u_i^2 \right) dx = \int_\Omega \varphi dx \quad \text{for all } \varphi \in C_0^\infty \left(\Omega \right)$$

become $\sigma_2 \left(D^2 u \right) = 1$, say, and $\triangle u > 0$, in the viscosity sense?

(The double divergence structure is readily seen from the well-known Gauss curvature formula for graph $(x_1, x_2, u(x)) \subset \mathbb{R}^3$ with induced metric $g : K = \left(-\frac{1}{2}\partial_{11}g_{22} + \partial_{12}g_{12} - \frac{1}{2}\partial_{22}g_{11}\right) / (\det g)^2 = \det D^2 u / \left(1 + |Du|^2\right)^2$.)

In dimension two, a (necessary) convexity condition should suffice, also for $\sigma_2\left(D^2 u\right) = 1$ in the equivalent Alexandrov sense. In general dimensions, what about $\triangle u > 0$ in the distribution sense for $u \in C^{1,2^+/3}$? The answer is yes in two dimensions, as shown by Pakzad ([P]). Moreover, in two dimensions, no better than $C^{1,1/3}$ "very weak" solution with sign changing $\triangle u$ have been constructed; see the work of Lewicka-Pakzad [LP] and [CS] [CHI]. It is worth noting that the singular solution to $\sigma_2\left(D^2 u\right) = 1$ constructed by C.Y. Li ([L]),

$$u(x) = \left(x_1^2 + \cdots + x_7^2\right) x_8^{7/5} - \frac{25}{84}x_8^{3/5} - \frac{25}{28}x_8^{14/5} \in W^{1,2}_{loc} \cap C^{3/5} \text{ jumps branches,}$$

because $\triangle u \approx -x_8^{-7/5} = \pm\infty$ near $x_8 = 0$.

References

[ALL]   Alvarez Olivier, Lasry Jean-Michel, Lions Pierre-Louis, *Convex viscosity solutions and state constraints*. J. Math. Pures Appl., (9) **76** (1997), no. 3, 265–288.

[BCGJ] Bao Jiguang, Chen Jingyi, Guan, Bo, Ji Min, *Liouville property and regularity of a Hessian quotient equation*. Amer. J. Math., **125** (2003), no. 2, 301–316.

[CGM]   Caffarelli Luis, Guan Pengfei, Ma Xi-Nan, *A constant rank theorem for solutions of fully nonlinear elliptic equations*. Comm. Pure Appl. Math., **60** (2007), no. 12, 1769–1791.

[CS]   Cao Wentao, Székelyhidi László Jr., *Very weak solutions to the two-dimensional Monge-Ampére equation*. Sci. China Math., **62** (2019), no. 6, 1041–1056.

[CHI]   Cao Wentao, Hirsch Jonas, Inauen, Dominik, $C^{1,1/3}$ *very weak solutions to the two dimensional Monge-Ampére equation*. arXiv:2310.06693.

[ChY]   Chang Sun-Yung Alice, Yuan Yu, *A Liouville problem for the sigma-2 equation*. Discrete Contin. Dyn. Syst., **28** (2010), no. 2, 659–664.

[CT]   Chaudhuri Nirmalendu, Trudinger Neil S., *An Alexsandrov type theorem for k-convex functions*. Bull. Austral. Math. Soc., **71** (2005), no. 2, 305–314.

[CHO]   Chen Chuanqiang, Han Fei, Ou Qiangzhong, *The interior C2 estimate for Monge-Ampere equation in dimension n=2*. Anal. PDE, **9** (2016), no. 6, 1419–1432.

[CSY]   Chen Jingyi, Shankar Ravi, Yuan Yu, *Regularity for convex viscosity solutions of special Lagrangian Equation*. Comm. Pure Appl. Math., **76** (2023), 4075–4086.

[CX]   Chen Li, Xiang Ni, *Rigidity theorems for the entire solutions of 2-Hessian equation*. J. Differential Equations, **267** (2019), no. 9, 5202–5219.

[CW]   Chou Kai-Seng, Wang Xu-Jia, *A variational theory of the Hessian equation.* Comm. Pure Appl. Math., **54** (2001), 1029–1064.

[GQ]   Guan Pengfei, Qiu Guohuan, *Interior C^2 regularity of convex solutions to prescribing scalar curvature equations.* Duke Math. J., **168** (2019), no. 9, 1641–1663.

[I]   Iwaniec Tadeusz, *On the concept of the weak Jacobian and Hessian.* Papers on analysis, Rep. Univ. Jyväskylä Dep. Math. Stat., **83** (2001), 181–205.

[LP]   Lewicka Marta, Pakzad Mohammad Reza, *Convex integration for the Monge-Ampère equation in two dimensions.* Anal. PDE, **10** (2017), no. 3, 695–727.

[L]   Li Caiyan, *Non-polynomial solutions to Hessian equations.* Calc. Var. Partial Differ. Equa., **60** (2021), no. 4, Paper No. 123, 6 pp.

[LT]   Lin Min, Trudinger Neil S., *On some inequalities for elementary symmetric functions.* Bull. Austral. Math. Soc., **50** (1994), 317–326.

[MSY]   McGonagle Matt, Song Chong, Yuan Yu, *Hessian estimates for convex solutions to quadratic Hessian equation.* Ann. Inst. H. Poincaré Anal. Non Linéire, **36** (2019), no. 2, 451–454.

[M]   Mooney Connor, *Strict 2-convexity of convex solutions to the quadratic Hessian equation.* Proc. Amer. Math. Soc., **149** (2021), no. 6, 2473–2477.

[N]   Nitsche Johannes C. C., *Elementary proof of Bernstein's theorem on minimal surfaces.* Ann. of Math., **66** (1957), 543–544.

[P]   Pakzad Mohammad Reza, *Convexity of weakly regular surfaces of distributional nonnegative intrinsic curvature.* arXiv:2206.09224.

[Q]   Qiu Guohuan, *Interior Hessian estimates for Sigma-2 equations in dimension three.* arXiv:1711.00948.

[S]   Savin Ovidiu, *Small perturbation solutions for elliptic equations.* Comm. Partial Differ. Equa., **32** (2007), no. 4-6, 557–578.

[SY1]   Shankar Ravi, Yuan Yu, *Hessian estimate for semiconvex solutions to the sigma-2 equation.* Calc. Var. Partial Differ. Equa., **59** (2020), no. 1, Paper No. 30, 12 pp.

[SY2]   Shankar Ravi, Yuan Yu, *Regularity for almost convex viscosity solutions of the sigma-2 equation.* J. Math. Study, **54** (2021), no. 2, 164–170.

[SY3]   Shankar Ravi, Yuan Yu, *Rigidity for general semiconvex entire solutions to the sigma-2 equation.* Duke Math. J., **171** (2022), no. 15, 3201–3214.

[SY4]   Shankar Ravi, Yuan Yu, *Hessian estimates for the sigma-2 equation in dimension four.* Submitted, arXiv:2305.12587.

[SY5]   Shankar Ravi, Yuan Yu, *Regularity for the Monge-Ampère equation by doubling.* preprint.

[T]   Trudinger Neil S., *Weak solutions of Hessian equations.* Comm. Partial Differ. Equa., **22** (1997), no. 7-8, 1251–1261.

[W]   Warren Micah, *Nonpolynomial entire solutions to σ_k equations.* Comm. Partial Differ. Equa., **41** (2016), no. 5, 848–853.

[WY]   Warren Micah, Yuan Yu, *Hessian estimates for the sigma-2 equation in dimension 3*. Comm. Pure Appl. Math., **62** (2009), no. 3, 305–321.

[X]   Xu Yingfeng, *Interior estimates for solutions of the quadratic Hessian equations in dimension three*. J. Math. Anal. Appl., **489** (2020), no. 2, 124179, 10 pp.

[Y]   Yuan Yu, *A Bernstein problem for special Lagrangian equations*. Invent. Math., **150** (2002), 117–125.

[Z]   Zhou Xingchen, *Notes on generalized special Lagrangian equation*. Submitted, 2023.

Transcendental Okounkov Bodies

KEWEI ZHANG

School of Mathematical Sciences, Beijing Normal University, Beijing, China

E-mail: kwzhang@bnu.edu.cn

Abstract In this survey, we first recall a classical construction in algebraic geometry, which associates a convex body to any big line on a projective manifold. Such convex body is called the Okunkov body, and it plays crucial roles in algebraic geometry. We then discuss how one can extend such construction to transcendental classes on a Kähler manifold.

In toric geometry, there is a striking correspondence between Delzant polytopes Δ and polarized toric manifolds (X_Δ, L_Δ). Much of the geometry of the toric manifold is encoded in its Delzant polytope; for instance, the volume of L_Δ, defined as

$$\operatorname{vol}(L_\Delta) = \lim_{k \to \infty} \frac{h^0(X_\Delta, kL_\Delta)}{k^n/n!},$$

is seen to equal $n!$ times the Euclidean volume of Δ.

Toric manifolds are very special, due to their large group of symmetries. Thus, it was quite remarkable that Okounkov in the 90s [24] showed how one can naturally associate a convex body $\Delta(L)$ to any polarized manifold (X, L). These bodies are now known as Okounkov bodies (or Newton–Okounkov bodies).

To construct the Okounkov body, one fixes a complete flag of submanifolds $Y_\bullet = (X = Y_0 \supset Y_1 \supset \cdots \supset Y_{n-1} \supset Y_n)$. To any $s \in H^0(X, kL) \setminus \{0\}$ we associate a valuation vector $\nu(s) = (\nu_1, \ldots, \nu_n) \in \mathbb{R}^n$ in the following way. First, we let ν_1 be the order of vanishing of s along Y_1. Next, if s_{Y_1} is a defining section for Y_1, we get that $(s/s_{Y_1}^{\nu_1})|_{Y_1}$ is a non-zero section of $L^k \otimes \mathcal{O}(-\nu_1 Y_1)$ restricted to Y_1. We set ν_2 to be the order of vanishing along Y_2 of this section.

Continuing this process yields the full valuation vector $\nu(s)$. The Okounkov body of L, denoted $\Delta_{Y_\bullet}(L)$, is then defined as the closure of the set of rescaled valuation vectors

$$\Delta_{Y_\bullet}(L) = \overline{\{\nu(s)/k : s \in H^0(X, kL) \setminus \{0\}, k \in \mathbb{N}\}}.$$

Convexity of the Okounkov body follows from the fact that $\nu(s \otimes t) = \nu(s) + \nu(t)$. A more difficult fact, however, is that the Okounkov body satisfies the volume identity:

$$\mathrm{vol}_{\mathbb{R}^n}(\Delta_{Y_\bullet}(L)) = \frac{1}{n!}\mathrm{vol}(L). \tag{1.1}$$

Later Lazarsfeld–Mustaţă [21] and Kaveh–Khovanskii [20] showed that Okounkov's construction works for more general big line bundles as well, together with the analogous volume identity (1.1). One should note that the volume of big line bundles is a much more subtle notion compared to the ample case; using convex bodies to study this object was indeed a breakthrough. For example, Lazarsfeld–Mustaţă used Okounkov bodies to show that the volume function is C^1 on the cone of big $\mathbb{R}$-divisors [21, Corollary C]. Since then, numerous applications of Okounkov bodies have been found in algebraic and complex geometry (see e.g. [2,5,9,18,22], to only name a few works).

The Okounkov body is a numerical invariant of the line bundle, i.e., it only depends on the first Chern class $c_1(L)$ of L, which is a real cohomology class of bidegree $(1, 1)$. Real cohomology classes that do not correspond to line bundles are called transcendental. The weaker positivity notions of being big or pseudoeffective have been extended to transcendental classes by Demailly, and Boucksom showed in [7] how to define the volume $\mathrm{vol}(\xi)$ of a transcendental class ξ.

In light of the above, given a flag $Y_\bullet$, a very natural question raised by Lazarsfeld–Mustaţă [21, p.831] asks whether one can define the Okounkov body $\Delta_{Y_\bullet}(\xi)$ associated with a transcendental class ξ, in such a way that it satisfies the volume identity

$$\mathrm{vol}_{\mathbb{R}^n}(\Delta_{Y_\bullet}(\xi)) = \frac{1}{n!}\mathrm{vol}(\xi). \tag{1.2}$$

There are some notable open problems related to the volume of big transcendental classes, such as obtaining transcendental holomorphic Morse inequalities, or the continuous differentiability of the volume on the big cone [3]. Identity (1.2) could turn out to be useful with tackling these problems, the same way as it was in the line bundle case [21, Corollary C].

As we explain now, in [14] Deng proposed a construction of Okounkov bodies $\Delta(\xi)$ for a transcendental class ξ. In the transcendental case global sections make no sense, but these classes admit a lot of currents. The vanishing order of a global section s of line bundle can be expressed using Lelong numbers of the current of integration $[s = 0]$. Deng's construction rests on the idea that computing Lelong numbers of appropriate currents should be the transcendental analogue of computing vanishing order of sections in the algebraic case.

Let (X, ω) be a connected compact Kähler manifold of dimension n. Let $\xi = \{\theta\} \in H^{1,1}(X, \mathbb{R})$ be a big (psef) cohomology class, with θ being a smooth closed $(1, 1)$-form. Assume that X has a complete flag of submanifolds

$$Y_\bullet = (X = Y_0 \supset Y_1 \supset \cdots \supset Y_{n-1} \supset Y_n),$$

Let $\mathcal{A}(X, \xi)$ be the set of positive currents $\theta + dd^c u$ in ξ with analytic singularities. This means that locally $u = c \log(\sum_j |f_j|^2) + h$, for $c \in \mathbb{Q}_+$, f_j holomorphic, and h bounded.

To any $T_0 \in \mathcal{A}(X, \xi)$ one associates a valuation vector $\nu(T_0) = (\nu_1, \ldots, \nu_2) \in \mathbb{R}^n$ in the following manner. Let ν_1 equal $\nu(T_0, Y_1)$, the Lelong number of T_0 along Y_1. One can show that $T_1 := (T_0 - \nu_1[Y_1])|_{Y_1}$ has analytic singularities on Y_1. Then one sets $\nu_2 := \nu(T_1, Y_2)$ and continues the above process until all components of $\nu(T_0)$ are defined. The Okounkov body of ξ, denoted $\Delta_{Y_\bullet}(\xi)$ (or just $\Delta(\xi)$, once the flag is fixed), is then defined as:

$$\Delta(\xi) = \overline{\nu(\mathcal{A}(X, \xi))} \subset \mathbb{R}^n.$$

As in the line bundle case, the convexity of the Okounkov body follows from the fact that $\nu(S + T) = \nu(S) + \nu(T)$. When ξ is a psef class, one puts $\Delta(\xi) = \cap_{\varepsilon > 0} \Delta(\xi + \varepsilon\{\omega\})$.

Deng's body coincides with Okounkov's when ξ is the first Chern class of a big line bundle [14]. Thus, this body is a natural generalization of the Okounkov body to the transcendental setting, and we will work with Deng's definition in our paper.

It remains to show that the transcendental Okounkov body satisfies (1.2).
Deng proved that this is true when the dimension of X is at most two, but the
general case was open until now. In our recent joint work [12] we obtain the
following result, which proves the volume identity in full generality:

Theorem 1   *Let ξ be a pseudoeffective cohomology class on X and $Y_\bullet$ be a
flag on X. Then the associated Okounkov body $\Delta(\xi)$ of ξ satisfies the volume
equality*

$$\mathrm{vol}_{\mathbb{R}^n}(\Delta(\xi)) = \frac{1}{n!}\,\mathrm{vol}(\xi). \tag{1.3}$$

Theorem 1 confirms [14, Conjecture 1.4] and also gives an affirmative answer
to the question of Lazarsfeld–Mustaţă [21, p.831].

We note that, *a priori* X might not contain any proper complex submanifolds
of positive dimension, let alone flags. For example, this happens if X is a non-
algebraic simple compact torus. However, one can always blow up X at a point
p. Then one can construct a flag $Y_\bullet$ on $\mathrm{Bl}_p X$ with $Y_1 \cong \mathbb{CP}^{n-1}$ being the
exceptional divisor and $Y_2 \supset \cdots \supset Y_n$ being a flag of linear subspaces in Y_1.
Therefore, up to modifying X and pulling back ξ, our main result is always
applicable.

The proof of Theorem 1 proceeds by induction on $\dim X$, with the recent
work of the third named author [29] helping with the induction step. In our
proof, it will be necessary to approximate the volume of $\Delta(\xi)$ using the volume
of partial Okounkov bodies, as studied in [30]. The advantage of our partial
Okounkov bodies is that they have good bimeromorphic properties, allowing to
reduce their study to the case when ξ is Kähler. In our proof we also need to
characterize those currents with analytic singularities whose pushforward also
has this property. This contrasts Deng's approach in the surface case [14], where
he can apply the Zariski decomposition, without modifying the surface (cf. [8]).

In our approach, an extension result slightly generalizing [10, Theorem
1.1] will be necessary to relate the slice volume of the Okounkov body to the
restricted volume of the big class. Let us state this extension result, which could
be of independent interest, in light of [16, Conjecture 37] and [11].

Theorem 2   *Let $V \subseteq X$ be a connected positive dimensional compact complex
submanifold of (X,ω), and $T = \omega|_V + \mathrm{dd}^c\varphi$ be a Kähler current on V. Assume*

that e^φ *is a Hölder continuous function on* V. *Then one can find a Kähler current* $\tilde{T} = \omega + \mathrm{dd}^c\tilde{\varphi}$ *on* X *such that* $\tilde{\varphi}|_V = \varphi$, *i.e.*, $\tilde{T}$ *extends* T. *Moreover,* $\tilde{\varphi}$ *is continuous on* $X \setminus V$ *and* $\mathrm{e}^{\tilde{\varphi}}$ *is Hölder continuous on* X. *In addition, if* φ *has analytic singularities, then so does* $\tilde{\varphi}$.

For the definition of analytic singularities we refer to [12, Definition 2.2]. Our definition is more general compared to [10]; see [12, Remark 2.7] for a comparison of various notions of analytic singularities appearing in the literature.

Potentials with exponential Hölder continuity often appear in the literature (see [4, 25]), so this result may find other applications in Kähler geometry. Similar to [10], we prove our extension result using Richberg's gluing technique. Note that, in our proof we used the exposition from [13, Section I.5.E], which allows for some simplifications.

Note that one can not expect Theorem 2 to hold for big classes, due to the obvious obstruction coming from the non-Kähler locus. In fact one can hope to extend a Kähler current in a big class only if it has appropriate prescribed singularities along the non-Kähler locus. This technical point inevitably arises in our proof of Theorem 1, and is circumvented with the help of partial Okounkov bodies.

In the end, we describe an alternative way of defining a convex body associated with a big class ξ (we refer the reader to [12, §8] for a more detailed construction). Instead of using the valuation vectors $\nu(T)$ of closed positive currents $T \in \xi$, we consider certain iterated toric degenerations of X, where in n steps (X, T) is degenerated to $(\mathbb{P}^n, T_n)$. Here, T is a closed positive $(1,1)$-current in ξ, while T_n is a closed positive and *toric* $(1,1)$-current on $\mathbb{P}^n$. As described in [12, §8], any such toric degeneration $(\mathbb{P}^n, T_n)$ comes with a naturally associated moment body $\Delta(T_n)$. This allows us to define the moment body $\Delta^\mu(\xi)$ of ξ as the closure of the union of the moment bodies of all such metrized toric degenerations of (X, ξ). We note that an algebraic version of this construction is considered in the recent thesis of Murata [23]. We show that this alternative body agrees with the Okounkov body of ξ:

Theorem 3 *For* ξ *big the moment body* $\Delta^\mu(\xi)$ *coincides with the Okounkov body* $\Delta(\xi)$.

As a direct consequence of Theorem 3 we get that $\Delta(T_n) \subseteq \Delta(\xi)$ for all toric degenerations $(\mathbb{P}^n, T_n)$ of (X, ξ), and that for any $\epsilon > 0$ we can find a toric degeneration $(\mathbb{P}^n, T_n)$ of (X, ξ) such that the Hausdorff distance between $\Delta(T_n)$ and $\Delta(\xi)$ is less than ϵ. This provides a strong link between transcendental Okounkov bodies and toric degenerations, which will be useful in further investigations.

The above result is inspired by the works of Harada–Kaveh [17] and Kaveh [19], who, building on the earlier work of Anderson [1], prove related results in the algebraic setting. This theorem is also related to work on canonical growth conditions [28].

Further directions Finally, we mention several appealing directions that one could explore further. One may try to define the Chebyshev transform of [27] in the transcendental setting to study weak geodesic rays in connection with transcendental notions of K-stability attached to cscK metrics [15, 26]. One could try to use Okounkov bodies to make progress on the conjectured transcendental Morse inequality [3], or to show differentiability of the volume in the big cone, akin to [21, Corollary C]. Moreover, one can study the mixed volume of Okounkov bodies and investigate its relation with the movable intersection product of big classes introduced in [6].

Regarding the extension of Kähler currents, it is intriguing to understand how far the Richberg type gluing techniques can be pushed.

The results in [12] seem to naturally extend to manifolds of the Fujiki class $\mathcal{C}$. Going beyond, we ask if one can attach Okounkov bodies to transcendental pseudoeffective cohomology classes ξ on general compact complex manifolds.

References

[1] D. Anderson. Okounkov bodies and toric degenerations. Math. Ann., 2013, 356(3): 1183-1202.

[2] H. Blum and M. Jonsson. Thresholds, valuations, and K-stability. Adv. Math., 2020, 365: 107062.

[3] S. Boucksom, J.-P. Demailly, M. Paun, et al. The pseudo-effective cone of a compact Kähler manifold and varieties of negative Kodaira dimension. J. Algebraic Geom., 2013, 22(2): 201-248.

[4]   S. Boucksom, C. Favre, and M. Jonsson. Valuations and plurisubharmonic singularities. Publ. Res. Inst. Math. Sci., 2008, 44(2): 449-494.

[5]   S. Boucksom, A. Küronya, C. Maclean, and T. Szemberg. Vanishing sequences and Okounkov bodies. Math. Ann., 2015, 361(3-4): 811-834.

[6]   S. Boucksom. Cônes positifs des variétés complexes compactes. PhD Thesis. Université Joseph-Fourier Grenoble, 2002.

[7]   S. Boucksom. On the volume of a line bundle. Internat. J. Math., 2002, 13(10): 1043-1063.

[8]   S. Boucksom. Divisorial Zariski decompositions on compact complex manifolds. Ann. Sci. Ecole Norm. Sup., 2004, 37(4): 45-76.

[9]   S. Boucksom and H. Chen. Okounkov bodies of filtered linear series. Compos. Math., 2011, 147(4): 1205-1229.

[10] T. C. Collins and V. Tosatti. An extension theorem for Kähler currents with analytic singularities. Ann. Fac. Sci. Toulouse Math., 2014, 23(4): 893-905.

[11] D. Coman, V. Guedj and A. Zeriahi. Extension of plurisubharmonic functions with growth control. J. Reine Angew. Math., 2013, 676: 33-49.

[12] T. Darvas, R. Reboulet, D. Witt Nystrom, M. Xia, K. Zhang, Transcendental Okounkov bodies. 2023 arXiv:2309.07584.

[13] J.-P. Demailly. Complex Analytic and Differential Geometry. Available on personal website, 2012.

[14] Y. Deng. Transcendental Morse inequality and generalized Okounkov bodies. Algebr. Geom., 2017, 4(2): 177-202.

[15] R. Dervan and J. Ross. K-stability for Kähler manifolds. Math. Res. Lett., 2017, 24(3): 689-739.

[16] S. Dinew, V. Guedj, and A. Zeriahi. Open problems in pluripotential theory. Complex Var. Elliptic Equ., 2016, 61(7): 902-930.

[17] M. Harada and K. Kaveh. Integrable systems, toric degenerations and Okounkov bodies. Invent. Math., 2015, 202(3): 927-985.

[18] C. Jiang and Z. Li. Algebraic reverse Khovanskii–Teissier inequality via Okounkov bodies. 2021. arXiv: 2112.02847.

[19] K. Kaveh. Toric degenerations and symplectic geometry of smooth projective varieties. J. London Math. Soc., 2019, 99(2): 377-402.

[20] K. Kaveh and A. G. Khovanskii. Newton-Okounkov bodies, semigroups of in- tegral points, graded algebras and intersection theory. Ann. of Math., 2012, 176(2): 925-978.

[21] R. Lazarsfeld and M. Mustata. Convex bodies associated to linear series. Ann. Sci. Ec. Norm. Super., 2009, 42(5): 783-835.

[22] C. Li and C. Xu. Stability of valuations: Higher rational rank. Peking Math. J., 2018, 1(1): 1-79.

[23] T. Murata. Toric degenerations of projective varieties with an application to equivariant Hilbert functions. Ph.D. Thesis. University of Pittsburgh. ProQuest LLC, Ann Arbor, MI, 2020.

[24] A. Okounkov. Brunn-Minkowski inequality for multiplicities. Invent. Math., 1996, 125(3): 405-411.

[25] J. Ross and D. Witt Nyström. Envelopes of positive metrics with prescribed singularities. Ann. Fac. Sci. Toulouse Math., 2017, 26(3): 687-728.

[26] Z. Sjöström Dyrefelt. K-semistability of cscK manifolds with transcendental cohomology class. J. Geom. Anal., 2018, 28(4): 2927-2960.

[27] D. Witt Nyström. Transforming metrics on a line bundle to the Okounkov body. Ann. Sci. Ec. Norm. Super., 2014, 47(6): 1111-1161.

[28] D. Witt Nyström. Canonical growth conditions associated to ample line bundles. Duke Math. J., 2018, 167(3): 449-495.

[29] D. Witt Nyström. Deformations of Kähler manifolds to normal bundles and restricted volumes of big classes. J. Differential Geom. (to appear). 2021. arXiv:2103.03660.

[30] M. Xia. Partial Okounkov bodies and Duistermaat-Heckman measures of non-Archimedean metrics. 2021. arXiv: 2112.04290.

Log Gradient Estimates for Heat Type Equations on Manifolds

QI ZHANG

Department of Mathematics, University of California, Riverside, CA 92521, USA

qizhang@math.ucr.edu

Abstract In this short survey paper, we first recall the log gradient estimates for the heat equation on manifolds by Li-Yau, R. Hamilton and later by Perelman in conjunction with the Ricci flow. Then we will discuss some of their applications and extensions focusing on sharp constants and improved curvature conditions.

1 Introduction

Gradient estimates for solutions of differential equations can be traced back at least to the study of analytic functions whose gradients in a smooth, bounded domain can be determined by their boundary values via the Cauchy integral formula. In the real variable case, Bernstein [Ber] discovered gradient bounds in a similar spirit for solutions of several elliptic equations in the interior of a domain. His method is to apply the maximum principle on a carefully chosen auxiliary function involving the solution and the modulus of the gradient, which satisfies a differential inequality whose nonlinear term has a good sign. This general strategy was very fruitful and is still being used nowadays for a wide range of equations and problems, some of which will be touched below. When the background space is a Riemannian manifold, the study of harmonic forms also involves gradient estimates in the L^2 and other integral sense. Hodge theory is one of the ramifications. Comparing with the Euclidean case, various curvature terms come into the play through commutation formulas such as Bochner's. This not only makes the situation more complicated but, more importantly, also reveals the connection with geometric and topological properties of the manifolds. Further applications of gradient estimates and the Bernstein technique in conjunction with commutation formulas are found in complex geometry such as Kodaira vanishing theorem and fully nonlinear elliptic equations which are beyond the scope of this paper.

2 Li-Yau and Hamilton estimates in fixed metric case

When the solution of an equation is positive. One can apply Bernstein's method to study the log gradient bound of the solution. This was carried out by S. T. Yau [Ya] and Cheng and Yau [CY] for positive harmonic functions on Riemannian manifolds. The log gradient bound is stronger than just the gradient bound. For example, log gradient bound for positive harmonic functions implies the Harnack inequality which roughly states that in a compact sub-domain bounded away from the boundary, the maximum value is bounded by the minimum value times a constant which is independent of the solution. See also [Mo] by Modica for a log gradient bound for nonlinear elliptic equations.

Log gradient bounds have been extended to positive solutions of the heat equation, which is the main topic of the present paper. Let $(\mathbf{M}^n, g_{ij})$ be a complete Riemannian manifold. In [LY], P. Li and S.T. Yau discovered the following celebrated Li-Yau bound.

Theorem 1 *Let* $\mathbf{M}$ *be a complete Riemannian n-manifold. Suppose* $Ric \geqslant -K$, *where* $K \geqslant 0$ *and* Ric *is the Ricci curvature of* $\mathbf{M}$. *Let* u *be a positive solution of the heat equation in* $\mathbf{M} \times (0, \infty)$:

$$\frac{\partial u}{\partial t} = \Delta u. \tag{2.1}$$

Then

$$\frac{|\nabla u|^2}{u^2} - \alpha \frac{u_t}{u} \leqslant \frac{n\alpha^2 K}{2(\alpha - 1)} + \frac{n\alpha^2}{2t}, \quad \forall \alpha > 1. \tag{2.2}$$

In the special case where $Ric \geqslant 0$, *then*

$$\frac{|\nabla u|^2}{u^2} - \frac{u_t}{u} \leqslant \frac{n}{2t}. \tag{2.3}$$

In the same paper, many applications of (2.2) and (2.3) have also been demonstrated by the authors, some of which are listed below.

- The classical parabolic Harnack inequality, Gaussian estimates of the heat kernel. We only state here the results when the Ricci curvature is nonnegative, which are sharp. General results can be found in the original paper by Li-Yau and the book [Lib].

Theorem 2 *Let* **M** *be a complete noncompact* n *dimensional Riemannian manifold of nonnegative Ricci curvature. Let* $u = u(x,t)$ *be a positive solution of the heat equation* $\Delta u - \partial_t u = 0$, $t > 0$. *Then*

$$u(x, t_1) \leqslant u(y, t_2)(t_2/t_1)^{n/2} \exp(d^2(x,y)/(4(t_2 - t_1)), \quad x, y \in \mathbf{M}, t_2 > t_1 > 0.$$

Theorem 3 *Let* **M** *be a complete noncompact Riemannian manifold of nonnegative Ricci curvature. Let* $G = G(x,t,y)$ *be the (minimal) heat kernel, i.e.* $\Delta_x G - \partial_t G = 0$, $t > 0$ *and* $G(x, 0, y) = \delta(x, y)$, *the Dirac delta function. For any small* $\epsilon > 0$, *there exists a positive constant* C, *depending only on* ϵ *and the dimension such that*

$$\frac{1}{C|B(x, \sqrt{t})|} \exp(-\frac{d^2(x,y)}{4(1-\epsilon)t}) \leqslant G(x, t, y) \leqslant \frac{C}{|B(x, \sqrt{t})|} \exp(-\frac{d^2(x,y)}{4(1+\epsilon)t})$$

Here $d = d(x, y)$ *is the geodesic distance between* x *and* y.

- Estimates of eigenvalues of the Laplace operator, and estimates of the Green's function of the Laplacian.
- Moreover (2.3) coupled with the small time asymptotic expansion of the heat kernel can even imply the Laplacian/volume Comparison Theorem (see e.g. [Chowetc] page 394).
- Later P. Li [Li] himself used the Harnack inequality to study the fundamental group of noncompact manifolds. He proved the following theorem.

Theorem 4 *Let* **M** *be a complete noncompact Riemannian manifold of nonnegative Ricci curvature. Suppose the volume of geodesic balls centered at* *a point is maximum in the sense that $|B(p,r)|/r^n \geqslant const. > 0$ for all $r > 0$. Then the fundamental group $\pi_1(\mathbf{M})$ is finite.*

This confirms a special case of Milnor's conjecture: *The fundamental group of complete noncompact Riemannian manifold of nonnegative Ricci curvature is finitely generated.* See also Anderson [An] for a different proof. Very recently Elia Brue, Aaron Naber, Daniele Semola posted a paper [BNS] to give a counter example. But this makes Li's result even more interesting and one can expect further generalization on the positive side of the result. Earlier positive results can be found in C. Sormani [So] (linear volume growth), G. Liu [Liu] (3 dimensions), see also J.Y. Pan [Pa] (3 dimensions).

• See also Mabuchi [Ma], Cao-Tian-Zhu [CTZ] for applications on complex geometry. These authors used Li-Yau inequality for certain (weighted) heat equation to prove the lower bound of the Green's function of certain perturbed Laplacian, which is used to study Kähler Einstein metrics and Kähler Ricci solitons.

We should mention that in the Euclidean space, the bound (2.3) was obtained by Aronson-Benilan [ArBe] earlier even for some porous medium equations. Hamilton [Ha93] further showed a matrix Li-Yau bound for the heat equation. This is the precursor to his matrix and trace Harnack inequality for 3 dimensional Ricci flows with positive curvature operator. Similar matrix Li-Yau bound was subsequently obtained by Cao-Ni [CaNi] on Kähler manifolds. We will discuss some of these results a little later.

For the past three decades, many Li-Yau type bounds have been proved not only for the heat equation, but more generally, for other linear and semi-linear parabolic equations on manifolds with or without weights. Let us mention the result by Bakry and Ledoux [BL] who derived the Li-Yau bound for weighted manifolds by an ordinary differential inequality involving the entropy and energy of the backward heat equation. Also P. Li-G. Tian [LT] derived a Li-Yau bound for certain algebraic varieties, J.Y. Li [Lij] proved mixed gradient estimates for the heat kernel. Davies [Dav], Garofalo-Modino [GM], J.F. Li and X. Xu [LX], Qian-Zhang-Zhu [QZZ], F.Y. Wang [Wan], Jiaping Wang [WanJ] and the latest Bakry-Bolley-Gentil [BBG] for improved coefficients and its references.

Next we turn to another important log gradient bound due to R. Hamilton

1993 [Ha93].

Theorem 5   *Let $u = u(x, t)$ be a positive solution to the heat equation $\Delta u - \partial_t u = 0$ on a compact manifold $\mathbf{M}$ whose Ricci curvature is bounded from below by a constant $-K \leqslant 0$. If $u \leqslant A$, a constant, then*

$$t\frac{|\nabla u|^2}{u} \leqslant (1 + 2Kt)u\ln\frac{A}{u}.$$

One reason for its importance is that it is a gradient estimate related to the "density" of the entropy $-u\ln u$. It is related to the second law of thermodynamics on the monotonicity of the entropy $-\int u\ln u\,dg$ and actually implies that the time derivative of the entropy is also monotone. So the entropy is concave in time if *Ricci* $\geqslant 0$.

Indeed, write

$$E(t) = -\int_{\mathbf{M}} u\ln u\,dg, \tag{2.4}$$

then using the basic identities

$$(\Delta - \partial_t)(u\ln u) = \frac{|\nabla u|^2}{u},$$

$$(\Delta - \partial_t)\frac{|\nabla u|^2}{u} = \frac{2}{u}\left|u_{ij} - \frac{u_i u_j}{u}\right|^2 + 2R_{ij}\frac{u_i u_j}{u},$$

one deduces

$$E'(t) = \int_{\mathbf{M}} \frac{|\nabla u|^2}{u}\,dg \geqslant 0, \tag{2.5}$$

$$E''(t) = -2\int_{\mathbf{M}} \frac{1}{u}\left|Hess\,u - \frac{\nabla u \otimes \nabla u}{u}\right|^2 dg - 2\int_{\mathbf{M}} \frac{Ric(\nabla u, \nabla u)}{u}\,dg \leqslant 0. \tag{2.6}$$

The proof of Theorem 5 is to use a linear combination of the basic identities and apply the maximum principle.

Another important application of Theorem 5 is that it allows comparison of values of the solution u at different spatial points in the same time level. This is not possible from the classical Harnack inequality. See further extensions in [ChH].

There is a localized version of Theorem 5, which has a square on top of the log term. It looks worse than Hamilton's global result but it is actually necessary.

Theorem 6 ([SZ]) *Let $\mathbf{M}$ be a Riemannian manifold of dimension $n \geqslant 2$ such that*

$$Ricci(\mathbf{M}) \geqslant -k, \qquad k \geqslant 0.$$

Suppose u is any positive solution to the heat equation in $Q_{R,T} \equiv B(x_0, R) \times [t_0 - T, t_0] \subset \mathbf{M} \times (-\infty, \infty)$. Suppose also $u \leqslant A$ in $Q_{R,T}$. Then there exists a dimensional constant c such that

$$\frac{|\nabla u(x,t)|}{u(x,t)} \leqslant c\Big(\frac{1}{R} + \frac{1}{T^{1/2}} + \sqrt{k}\Big)\Big(1 + \ln\frac{A}{u(x,t)}\Big) \qquad (2.7)$$

in $Q_{R/2, T/2}$.

Parabolic Liouville theorem for ancient solutions of the heat equation is one of the applications of this localized gradient estimate. See [LinZ], [Mos] and [HY] for further development and refinement in this direction.

3 Perelman type estimates in Ricci flow case

The study of Li-Yau bound for heat type equations under the Ricci flow was initiated by Hamilton. In [Ha4], he obtained a Li-Yau bound for the scalar curvature along the Ricci flow on 2-sphere. This result was later improved by Chow [Chow]. In higher dimensions, both matrix and trace Li-Yau bounds for curvature tensors, also known as Li-Yau-Hamilton inequalities, were obtained by Hamilton [Ha5] for the Ricci flow with bounded curvature and nonnegative curvature operator. These estimates played a crucial role in the study of singularity formations of the Ricci flow on three-manifolds and the solution to Poincaré conjecture. We remark that Brendle [Bre] has generalized Li-Yau-Hamilton inequalities under weaker curvature assumptions. The Li-Yau-Hamilton inequality for the Kähler-Ricci flow with nonnegative holomorphic bisectional curvature was obtained by H.-D. Cao [Cao].

On the other hand, the log gradient bounds in Section 2 have been extended to situations with moving metrics. Let $g_{ij}(t)$, $t \in [0, T]$, be a family of Riemannian metrics on $\mathbf{M}$ which solves the Ricci flow:

$$\frac{\partial}{\partial t} g_{ij}(t) = -2R_{ij}(t), \qquad (3.1)$$

where $R_{ij}(t)$ is the Ricci curvature tensor of $g_{ij}(t)$. One may still consider linear and semi-linear parabolic equations under the Ricci flow in the sense that in the heat operator $\dfrac{\partial}{\partial t} - \Delta$, we have $\Delta = \Delta_t$ which is the Laplace operator with respect to the metric $g_{ij}(t)$ at time t.

The two most prominent examples are the heat equation

$$(\Delta - \frac{\partial}{\partial t})u = 0, \ \partial_t g_{ij} = -2R_{ij} \tag{3.2}$$

and the conjugate heat equation

$$H^*u \equiv (\Delta - R + \frac{\partial}{\partial t})u = 0, \quad \partial_t g_{ij} = -2R_{ij}. \tag{3.3}$$

In 2002, in the breakthrough paper [P1], Perelman established the equivalent of (2.5) and (2.6) for (3.3). More precisely,

Definition 7 (F entropy and W entropy) Let u be a positive solution to (3.3) in a compact n-manifold. The F entropy is the integration of $H^*(u\ln u)$, i.e.

$$\mathbf{F} = \int_{\mathbf{M}} (\frac{|\nabla u|^2}{u} + Ru)dg(t). \tag{3.4}$$

Here R is the scalar curvature. The W entropy is a combination of the F entropy and the Boltzmann entropy $\mathbf{B} = \displaystyle\int u \ln u \, dg(t)$ together with certain scaling factor. Let $\tau > 0$ be such that $\dfrac{d\tau}{dt} = -1$, define

$$\mathbf{W} = \tau \mathbf{F} - \mathbf{B} - \frac{n}{2}(\ln 4\pi\tau) - n. \tag{3.5}$$

i.e.

$$\mathbf{W} = \int_{\mathbf{M}} \left[\tau(\frac{|\nabla u|^2}{u} + Ru) - u\ln u - \frac{n}{2}(\ln 4\pi\tau)\, u - nu\right] dg(t).$$

Theorem 8 ([P1]) *Perelman's F and W entropy are nondecreasing in time t. Moreover*

$$\frac{d}{dt}\mathbf{F} = 2\int_{\mathbf{M}} |Ric - Hess(\ln u)|^2 \, u \, dg(t),$$

$$\frac{d}{dt}\mathbf{W} = 2\tau \int_{\mathbf{M}} \left[Ric - Hess(\ln u) - \frac{1}{2\tau}g\right]^2 u \, dg(t).$$

The amazing thing is that F and W are monotone in time without any curvature condition. The monotonicity is strict unless the Ricci flow is gradient soliton. The proof of the theorem can be based on the following identities

$$H^*(u \ln u) = \frac{|\nabla u|^2}{u} + Ru,$$

$$H^*(\frac{|\nabla u|^2}{u} + Ru)$$

$$= \frac{2}{u}\left(u_{ij} - \frac{u_i u_j}{u}\right)^2 + 4\nabla R\nabla u + \frac{4}{u}Ric(\nabla u, \nabla u) + 2|Ric|^2 u + 2u\Delta R.$$

All curvatures are integrated out.

In [P1], Perelman also showed a combination of Li-Yau and Hamilton type bound for $u = u(x,\tau;y,0)$ the fundamental solution of the conjugate heat equation (3.3) with a pole at $(y,0)$ under the Ricci flow on a compact manifold: for $\tau = T - t$ with T a fixed number,

$$\tau\left(-2\frac{\Delta u}{u} + \frac{|\nabla u|^2}{u^2} + R\right) - \ln u - \frac{n}{2}(\ln 4\pi\tau) - n \leqslant 0.$$

This inequality implies:

- Perelman's monotonicity formula for the W entropy and,
- his κ non-collapsing property for volume of geodesic balls (on his way to the Poincaré conjecture);
- and after a little work, the non-inflating property for the volume of geodesic balls ([Z12], [CW13]).

These are some of the basic properties of Ricci flows by now.

The last two statements say that if the scalar curvature is bounded in a suitable region then

$$K_1 \leqslant |B(x,r,t)|/r^n \leqslant K_2, \qquad 0 < r \leqslant r_0, 0 < t \leqslant t_0,$$

for some positive constants K_1, K_2 which may depend on t_0 and the initial metric. Here $|B(x,r,t)|$ is the volume of the geodesic ball of radius r centered at x with respect to the metric at time t. r_0 is a fixed scale. Note that no Ricci curvature bound is needed.

Afterwards, there have been many results on Li-Yau bounds for positive solutions of the heat or conjugate heat equations under the Ricci flow.

- For example authors of Kuang-Zhang [KuZh] and X.D. Cao [Cx] proved (non-sharp) Li-Yau type bound for all positive solutions of the conjugate heat equation without any curvature condition, just like Perelman's aforementioned result for the fundamental solution.
- In X.D.Cao-Hamilton [CH] and Bailesteanu-Cao-Pulemotov, [BCP] the authors proved various Li-Yau type bounds for positive solutions of (3.2) under either positivity condition of the curvature tensor or boundedness of the Ricci curvature.

So there is a marked difference between these results on the conjugate heat equation and the heat equation in the curvature conditions. In view of the absence of curvature condition for the conjugate heat equation, one would hope that the curvature conditions for the heat equation can be weakened.

A few years ago, in Bamler-Zhang [BZ], the authors proved the following gradient estimate for bounded positive solutions u of the heat equation (3.2),

$$|\Delta u| + \frac{|\nabla u|^2}{u} - aR \leqslant \frac{Ba}{t}, \qquad (3.6)$$

where $R = R(x,t)$ is the scalar curvature of the manifold at time t, and B is a constant and a is an upper bound of u on $\mathbf{M} \times [0,T]$. Although this result requires no curvature condition and it has some other applications such as distance distortion estimate under Ricci flows, it is not a Li-Yau type bound. It turns out that a Li-Yau type bound is valid for the forward heat equation (3.2) under the very weak assumption that the scalar curvature is bounded.

Theorem 9 ([ZZ1]) *Let* $\mathbf{M}$ *be a compact* n *dimensional Riemmannian manifold, and* $g_{ij}(t)$, $t \in [0,T)$, *a solution of the Ricci flow* (3.1) *on* $\mathbf{M}$. *Denote by* R *the scalar curvature of* $\mathbf{M}$ *at* t, *and* R_1 *a positive constant. Suppose that* $-1 \leqslant R \leqslant R_1$ *for all time* t, *and* u *is a positive solution of the heat equation* (3.2). *Then, for any* $\delta \in [\frac{1}{2}, 1)$, *we have*

$$\delta \frac{|\nabla u|^2}{u^2} - \frac{\partial_t u}{u} \leqslant \delta \frac{|\nabla u|^2}{u^2} - \frac{\partial_t u}{u} - \alpha R + \frac{\beta}{R+2} \leqslant \frac{1}{t} \left(\frac{n}{2\delta} + \frac{4n\beta T}{\delta(1-\delta)} \right) \qquad (3.7)$$

for $t \in (0,T)$, *where* $\alpha = \dfrac{n}{2\delta(1-\delta)^2}$ *and* $\beta = \alpha(R_1 + 2)^2$.

Note that the curvature assumption is made only on the scalar curvature rather than on the Ricci or curvature tensor. In this sense, this assumption is essentially optimal. Under suitable assumptions, the result in the theorem still holds when **M** is complete noncompact, or with a little weakening of the scalar curvature condition. For the Ricci flow on a compact manifold **M**, one can always rescale a solution so that the scalar curvature is bounded from below by -1.

This theorem clearly implies a Harnack inequality for positive solutions of (3.2) if the scalar curvature is bounded.

In the proof, the main tool is still the maximum principle applied on a differential inequality involving Li-Yau type quantity. However, due to the Ricci flow, extra terms involving the Ricci curvature and Hessian of the solution will appear. In order to proceed we need to create a new term with the scalar curvature in the denominator. The key identity is the following. Let

$$F = -\Delta u + \delta \frac{|\nabla u|^2}{u} - \alpha Ru + \frac{\beta u}{R+C}, \tag{3.8}$$

and operator $\mathcal{L} = \Delta - \dfrac{\partial}{\partial t}$, where δ, α, and β are arbitrary constants and C is a constant so that $R + C > 0$. Then

$$\mathcal{L}F = \frac{1}{u}\left|u_{ij} - \frac{u_i u_j}{u} + uR_{ij}\right|^2 + \frac{2\delta - 1}{u}\left|u_{ij} - \frac{u_i u_j}{u}\right|^2$$

$$+ \frac{1}{(2\alpha - 1)u}\left|(2\alpha - 1)uR_{ij} + \frac{u_i u_j}{u}\right|^2$$

$$- \frac{1}{2\alpha - 1}\frac{|\nabla u|^4}{u^3} + \frac{\alpha u}{R+C}\left|\nabla R - \frac{R+C}{u}\nabla u\right|^2 \tag{3.9}$$

$$+ \left(\frac{\beta}{(R+C)^3} - \frac{\alpha}{R+C}\right)u|\nabla R|^2 - \frac{\alpha(R+C)|\nabla u|^2}{u}$$

$$+ \frac{2\beta u|R_{ij}|^2}{(R+C)^2} + \frac{\beta u}{R+C}\left|\frac{\nabla R}{R+C} - \frac{\nabla u}{u}\right|^2 - \frac{\beta|\nabla u|^2}{u(R+C)}.$$

Multiplying this by t and using the maximum principle yields the claimed result.

4 Li-Yau gradient bound under integral curvature condition

In all of these log gradient bounds in the static metric cases, the essential assumption is that the Ricci curvature or the corresponding Bakry-Emery Ricci curvature is bounded from below by a constant. In many situations, it is highly desirable to weaken this assumption. For example, in the case of Kähler Ricci flow in the Fano case, at each time slice, one does not know if the Ricci curvature is bounded. One only knows that $|Ric|$ is in L^4 by the work of G. Tian and Z. L. Zhang [TZz] and $|Ric|^2$ is in a Kato class by G. Tian and Q.S. Zhang [TZq2].

By the work of Petersen and Wei [PW1] and [PW2], we know that the classical volume comparison theorem can be generalized to the case with L^p curvature bound with $p > n/2$. If one also has Li-Yau bound, then one knows that the Poincaré and Harnack inequalities also hold. This allows one to extend, to certain extent, the Cheeger-Colding theory to the case of integral Ricci bound.

A few years ago, Meng Zhu and the author proved Li-Yau bounds for positive solutions for both the fixed metric case (2.1) and the Ricci flow case (3.2) under essentially optimal curvature conditions. The latter was discussed in the last section.

For the fixed metric case, we will have two independent conditions and two conclusions. The conditions are motivated by different problems such as studying manifolds with integral Ricci curvature bound and the Kähler-Ricci flow. The conclusions range from long time bound with necessarily worse constants, to short time bound with better constants.

The theorem basically says:

(a) $|Ric^-|$ in L^p, $p > n/2$ and volume noncollapsing $\Rightarrow$ Li-Yau inequality $\Rightarrow$ classical parabolic Harnack inequality. i.e. if u is a positive solution of the heat equation and $Q^- = B(x,r) \times [T_1, T_2]$ and $Q^+ = B(x,r) \times [T_3, T_4]$ are two cylinders in the domain of u such that $T_2 < T_3$. Then

$$\sup_{Q^-} u \leqslant C \inf_{Q^+} u.$$

Here C is a positive constant independent of u.

(b) Along each time slice of normalized Kähler Ricci flow, the Li-Yau inequality holds.

Theorem 10 ([ZZ1])   *Let* $(\mathbf{M}, g_{ij})$ *be a compact n dimensional Riemannian manifold, and u a positive solution of (2.1). Suppose either one of the following conditions holds.*

(a) $\displaystyle\int_{\mathbf{M}} |Ric^-|^p dy \equiv \sigma < \infty$ *for some* $p > \dfrac{n}{2}$, *where* Ric^- *denotes the nonpositive part of the Ricci curvature; and the manifold is noncollapsed under scale 1, i.e.,* $|B(x,r)| \geqslant \rho r^n$ *for* $0 < r \leqslant 1$ *and some* $\rho > 0$;

(b) $\displaystyle\sup_{\mathbf{M}} \int_{\mathbf{M}} |Ric^-|^2 d^{-(n-2)}(x,y) dy \equiv \sigma < \infty$ *and the heat kernel of (2.1) satisfies the Gaussian upper bound (which holds automatically under (a)):*

$$G(x,t;y,0) \leqslant \frac{\hat{C}(t)}{t^{n/2}} e^{-\bar{c}d^2(x,y)/t}, \ t \in (0,\infty) \tag{4.1}$$

for some positive constant $\bar{c}$ and positive increasing function $\hat{C}(t)$ which grows to infinity as $t \to \infty$.

Then, for any constant $\alpha \in (0,1)$, we have

$$\alpha \underline{J}(t) \frac{|\nabla u|^2}{u^2} - \frac{\partial_t u}{u} \leqslant \frac{n}{(2-\delta)\alpha \underline{J}(t)} \frac{1}{t} \tag{4.2}$$

for $t \in (0,\infty)$, where

$$\delta = \frac{2(1-\alpha)^2}{n + (1-\alpha)^2},$$

and

$$\underline{J}(t) = \begin{cases} 2^{-1/(5\delta^{-1}-1)} e^{-(5\delta^{-1}-1)\frac{n}{2p-n}} \left[4\sigma \hat{C}(t)^{1/p}\right]^{\frac{2p}{2p-n}} t, & (a); \\[2mm] 2^{-1/(10\delta^{-1}-2)} e^{-C_0(5\delta^{-1}-1)\sigma \hat{C}(t)t}, & (b), \end{cases} \tag{4.3}$$

with C_0 being a constant depending only on n, p and ρ, and $\hat{C}(t)$ the increasing function on the right hand side of (4.1).

In particular, for any $\beta \in (0,1)$, there is a $T_0 = T_0(\beta, \sigma, p, n, \rho)$ such that

$$\beta \frac{|\nabla u|^2}{u^2} - \frac{\partial_t u}{u} \leqslant \frac{n}{2\beta} \frac{1}{t} \tag{4.4}$$

for $t \in (0, T_0]$. Here $T_0 = c(1-\beta)^{4p/(2p-n)}$ and $c(1-\beta)^4$ under conditions (a) and (b), respectively; and c is a positive constant depending only on the parameters of conditions (a) and (b), i.e., $c = c(\sigma, p, n, \rho)$.

Are the conditions sharp? It turns out that the L^p condition on $|Ric^-|$ is almost necessary for solutions to be differentiable. For example one can not take $p = n/2$. The power 2 on top of $|Ric^-|$ in condition (b) can be replaced by 1 and the proof is identical. It is also realized in [Ro] and [Car] that a variation of this integral condition on $|Ric^-|$, using the heat kernel and space time integration, instead of $1/d(x,y)^{n-2}$ and space integral, which is also called as the Kato condition, essentially implies heat kernel upper bound with a growing parameter. Moreover the volume noncollasping condition can be removed.

For some constants p, $r > 0$, define $k(p,r) = \sup\limits_{x \in M} r^2 \left(\fint_{B(x,r)} |Ric^-|^p dV \right)^{1/p}$,

where Ric^- denotes the negative part of the Ricci curvature tensor. We prove that for any $p > \dfrac{n}{2}$, when $k(p,1)$ is small enough, certain Li-Yau type gradient bound holds for the positive solutions of the heat equation on geodesic balls $B(O,r)$ in $\mathbf{M}$ with $0 < r \leqslant 1$. Here the assumption that $k(p,1)$ being small allows the situation where the manifolds are collapsing.

Theorem 11 ([ZZ2]) *Let* $(\mathbf{M}^n, g_{ij})$ *be a complete Riemannian manifold and* u *a positive solution of the heat equation on* $\mathbf{M}$. *For any* $p > \dfrac{n}{2}$, *there exists a constant* $\kappa = \kappa(n,p)$ *such that the following holds. If* $k(p,1) \leqslant \kappa$, *then for any point* $O \in \mathbf{M}$ *and constant* $0 < \alpha < 1$, *we have*

$$\alpha \underline{J} \frac{|\nabla u|^2}{u^2} - \frac{\partial_t u}{u} \leqslant \frac{n}{\alpha(2-\delta)\underline{J}} \frac{1}{t} + \frac{C}{\alpha(2-\delta)\underline{J}} \left[\frac{1}{\alpha(2-\delta)\underline{J}(1-\alpha)} + 1 \right], \quad (4.5)$$

in $B(O, \dfrac{1}{2}) \times (0, \infty)$, *where*

$$\underline{J} = \underline{J}(t) = 2^{-\frac{1}{a-1}} \exp\left\{ -2C\kappa \left(1 + [2C(a-1)\kappa]^{\frac{n}{2p-n}} \right) t \right\},$$

$\delta = \dfrac{2(1-\alpha)^2}{n + (1-\alpha)^2}$, $a = \dfrac{5[n + (1-\alpha)^2]}{2(1-\alpha)^2}$ *and* $C = C(n,p)$ *is a constant depending on* n *and* p.

The theorem basically says if $K(p,r)$ is small then Li-Yau inequality holds. The smallness condition is necessary due to the example by Dean Yang [Yan], which is a dumb bell shaped manifold where $K(p,r)$ is not small. There is no volume doubling property for such a manifold. But we know Li-Yau inequality implies, among several things, the volume doubling property. See the discussion in a recent paper by X.Z. Dai, G.F. Wei and Z. L. Zhang [DWZ].

So the condition of this theorem is also essentially sharp.

In particular, on such manifolds the L^2 Poincaré inequality holds. It has been an open question for a while if this is true for collapsing manifolds with integral Ricci bound.

One can summarize the result as: $K(p, r)$ small implies:

(a) Harnack inequality for heat equation;

(b) volume doubling property and Poicaré inequality.

Notice that the only condition is on the Ricci curvature in terms of $K(p, r)$. Previously by the work of C. Croke [Cr], one knows that the pointwise bound $Ricci \geqslant -k$ implies the isoperimetric inequality and hence the L^2 Sobolev inequality. In [DWZ], the authors gave a direct proof of the isoperimetric inequality under integral Ricci bound using relative volume comparison and an idea by Gromov [Gro]. See also earlier result in [Gal].

From the work of Saloff-Coste [Sa] and Grigoryan [Gri], we know that from the Harnack inequality, there are a number of other implications such as Gaussian upper bound of heat kernel and the L^2 Sobolev inequality. By the work of Petersen and Wei [PW1], we also know it implies a number of geometric and topological results such as finiteness of diffeomorphism type.

Let us outline the idea of the proof. We will use the maximum principle on a quantity involving the solution and its gradient in the end. However, unlike the case with Ricci curvature bounded from below, the term involving the Ricci curvature can not be thrown away. The idea is to construct an auxiliary function by solving a nonlinear equation to cancel the term involving the Ricci curvature. The function is now referred to as the J function.

Let $J = J(x, t)$ be a smooth positive function and $\alpha \in (0, 1)$ be a parameter. Write

$$Q \equiv \alpha J \frac{|\nabla u|^2}{u^2} - \frac{\partial_t u}{u}. \tag{4.6}$$

and $f = \ln u$. After some computation, one finds that

$$(\Delta - \partial_t)(tQ) + 2\nabla \ln u \nabla(tQ) \geqslant \alpha t(2J - \delta J)\frac{1}{n}\left(|\nabla f|^2 - \partial_t f\right)^2$$

$$+ \alpha\left[\Delta J - 2VJ - 5\delta^{-1}\frac{|\nabla J|^2}{J} - \partial_t J\right]t|\nabla f|^2 - \delta \alpha t J|\nabla f|^4 - Q, \tag{4.7}$$

where we have written $|Ric^-| = V$ and $\delta > 0$ is another small parameter.

For any given parameter $\delta > 0$ such that $5\delta^{-1} > 1$, the problem

$$\begin{cases} \Delta J - 2VJ - 5\delta^{-1}\dfrac{|\nabla J|^2}{J} - \partial_t J = 0, \quad \text{on} \quad \mathbf{M} \times (0, \infty); \\ J(\cdot, 0) = 1, \end{cases} \tag{4.8}$$

has a unique solution for $t \in [0, \infty)$, which satisfies

$$\underline{J}(t) \leqslant J(x, t) \leqslant 1, \tag{4.9}$$

where

$$\underline{J}(t) = \begin{cases} 2^{-1/(5\delta^{-1}-1)} e^{-(5\delta^{-1}-1)\frac{n}{2p-n}} \left[4\sigma\hat{C}(t)^{1/p}\right]^{\frac{2p}{2p-n}t}, & \text{under condition (a)}; \\ 2^{-1/(10\delta^{-1}-2)} e^{-C_0(5\delta^{-1}-1)\sigma\hat{C}(t)t}, & \text{under condition (b)}, \end{cases} \tag{4.10}$$

with C_0 being a constant depending only on n, p and ρ, and $\hat{C}(t)$ the increasing function in (4.1). The term involving the Ricci curvature on (4.7) is cancelled. The maximum principle then implies the result in the theorem.

5   Li-Yau gradient bound with sharp constants

Except for the case where Ricci is nonnegative and Perelman's gradient bounds for the fundamental solution, these Li-Yau type bounds are not sharp. The difference between (2.3) and (2.2) is more than a formality since the former is actually a sharp Log Laplace estimate because it says

$$-\Delta \ln u = \frac{|\nabla u|^2}{u^2} - \frac{u_t}{u} \leqslant \frac{n}{2t}.$$

The Li-Yau bound (2.2) when Ricci curvature changes sign is not sharp.

The bound (2.2) was later improved for small time by Hamilton in [Ha93], where he proved under the same assumptions as above that

$$\frac{|\nabla u|^2}{u^2} - e^{2Kt}\frac{u_t}{u} \leqslant e^{4Kt}\frac{n}{2t}. \tag{5.1}$$

See also the work by Davies [Dav] and Yau himself [Yau94].

For the original Li-Yau inequality, in over 30 years, several sharpening of the bounds have been obtained with α replaced by several functions $\alpha = \alpha(t) > 1$ but not equal to 1. We have mentioned some of these works in Section 2. A well known open question (in Chow-Lu-Ni's book, [LX], [BBG] and several other places) asks if a sharp bound can be reached. In a recent short paper, we show that for all complete compact manifolds one can take $\alpha = 1$. Thus a sharp bound, up to computable constants, is found in the compact case. This result also seems to sharpen Theorem 1.4 in [LY] for compact manifolds with convex boundaries.

In the noncompact case, one can not take $\alpha = 1$ even for the hyperbolic space. An example is also given, which shows that there does not exist an optimal function $\alpha = \alpha(t, K)$ for all noncompact manifolds with negative Ricci lower bound $-K$, giving a negative answer to the open question in the noncompact case.

Theorem 12 ([Z21]) *Let $(\mathbf{M}, g_{ij})$ be a complete, compact n dimensional Riemannian manifold and u a positive solution of the heat equation on $\mathbf{M} \times (0, \infty)$, i.e.,*

$$(\Delta - \partial_t)u = 0. \tag{5.2}$$

Let $diam_{\mathbf{M}}$ be the diameter of $\mathbf{M}$ and suppose the Ricci curvature is bounded from below by a non-positive constant $-K$, i.e. $R_{ij} \geqslant -Kg_{ij}$. Then there exist dimensional constants $C_1 > 0$ and $C_2 > 0$ such that

$$-t\Delta \ln u = t\left(\frac{|\nabla u|^2}{u^2} - \frac{\partial_t u}{u}\right)$$
$$\leqslant \left(\frac{n}{2} + \sqrt{2nK(1+Kt)(1+t)}\, diam_{\mathbf{M}} + \sqrt{K(1+Kt)(C_1 + C_2K)t}\right). \tag{5.3}$$

The constants C_1 and C_2 arise from the standard volume comparison or/and heat kernel upper and lower bound which can be efficiently estimated. For compact manifolds, the main utility of the result is for small or finite time. For example, when we choose C_1 and C_2 to be independent of the volume comparison theorem, the bound actually gives rise to a Laplace and hence volume comparison theorem as $t \to 0$. For large time, a solution converges to a constant. As $t \to \infty$, it is expected that the bound will be of order $o(t)$

instead of $O(t)$ as of now. But we will not pursue the improvement this time. It would still be interesting to find a sharp form of the RHS of (5.3).

For the noncompact case, we give a negative answer on the existence of a sharp function which depends on time t and Ricci lower bound only and for which the Li-Yau bound holds. First, in the noncompact case one can not prove a bound like (5.3) with a finite right hand side for a fixed t even for the hyperbolic space. For example in the 3 dimensional hyperbolic space with the standard metric, the heat kernel (c.f. [Dav] e.g.) is:

$$G(x,t,y) = \frac{1}{(4\pi t)^{3/2}} \frac{d(x,y)}{\sinh d(x,y)} e^{-t - \frac{d^2(x,y)}{4t}}.$$

For $r = d(x,y)$,

$$-t\Delta \ln G(x,t,y)$$

$$= -t\sinh^{-2} r \frac{\partial}{\partial r} \left(\sinh^2 r \frac{\partial}{\partial r} \right) \ln \left(\frac{1}{(4\pi t)^{3/2}} \frac{r}{\sinh r} e^{-t - \frac{r^2}{4t}} \right)$$

is of the order $d(x,y)$ when $d(x,y) \to \infty$.

Therefore one can not take $\alpha = 1$ for all noncompact manifolds with Ricci bounded from below. Second, let $\mathbf{M}_0$ be any compact manifold of dimension n and $\mathbf{M} = \mathbf{M}_0 \times \mathbf{R}^1$ be the product manifold of $\mathbf{M}_0$ with $\mathbf{R}^1$. Then the heat kernel on $\mathbf{M}$ is given by $G = G_0(x,t,y)G_1(z,t,w)$ where G_0 and G_1 are the heat kernel on $\mathbf{M}_0$ and $\mathbf{R}^1$ respectively with $x, y \in \mathbf{M}_0$ and $z, w \in \mathbf{R}^1$ and $t > 0$. Fixing y and w, we compute

$$-t\Delta \ln G = -t(\Delta_0 + \partial_z^2) \ln G = -t\Delta_0 \ln G_0 - t\partial_z^2 \ln G_1$$

$$\leqslant \left(\frac{n+1}{2} + \sqrt{2nK(1+Kt)(1+t)} \, diam_{\mathbf{M}_0} + \sqrt{K(1+Kt)(C_1 + C_2 K)t} \right).$$

Here we just used the above Theorem on the compact manifold $\mathbf{M}_0$ with $-K$ being the Ricci lower bound and Δ_0 being the Laplace-Beltrami operator on $\mathbf{M}_0$. So by a result of C.J. Yu and F.F. Zhao [YZ] or [LWH+] , for the product manifold $\mathbf{M}$, we can take $\alpha = 1$ for all positive solutions, which is optimal.

These two examples show that there does not exist a single optimal function of time t and K only, which works for all noncompact manifolds with Ricci curvature bounded from below by a negative constant $-K$.

The reason is that for each negative constant $-K$, the optimal function for the above manifold $\mathbf{M}_0 \times \mathbf{R}^1$ is $\alpha = 1$. But for the hyperbolic space with $Ricci = -Kg$, the optimal function is worse than $\alpha = 1$.

The idea of the proof is the following. First we prove the sharp bound for heat kernels G instead of general positive solutions. This involves an iterated integral estimate on $-\Delta \ln G$ which satisfies a nonlinear heat type equation and Hamilton's estimate of $|\nabla \ln G|$ by $t \ln A/G$, taking advantage of the finiteness of the diameter of $\mathbf{M}$ and lower bound of G. Here A is a constant which dominates the heat kernel in a suitable time interval. Note that we do not follow the usual way of using the maximum principle. Next we use crucially the result in Yu-Zhao [YZ] which states that a Li-Yau type bound on the heat kernel implies the same bound on all positive solutions.

Recently an improvement is made by Xingyu Song, Ling Wu, and Meng Zhu on the parameters, resulting in some refinement and extension of the above result, Here is a sample in the Ricci flow case. See also [SW] for another extension.

Theorem 13 ([SWZ]) *Let M^n be an n-dimensional closed Riemannian manifold and $(M^n, g(x,t))_{t \in [0,2)}$ a solution to the Ricci flow $\dfrac{\partial}{\partial t} g(x,t) = -2Ric(x,t)$, $(x,t) \in M \times [0,2)$. Assume that for some constant $K \geqslant 0$, $-\dfrac{K}{2-t} g(x,t) \leqslant Ric(x,t) \leqslant \dfrac{K}{2-t} g(x,t)$ and $\nu[g_0, 4] \geqslant -nK$, $(x,t) \in M \times [0,2)$. Let u be a smooth positive solution of the heat equation $\left(\Delta_t - \dfrac{\partial}{\partial t} \right) u(x,t) = 0$ on $M \times [0,2)$ and $d_t(M)$ the diameter of $(M^n, g(x,t))$. Then there exists a constant c_9 depending on n, K such that*

$$t \left(\frac{|\nabla u|^2}{u^2} - \frac{\partial_t u}{u} \right) \leqslant \frac{n}{2} + \frac{nKt}{2-t} + \sqrt{\frac{Kc_9}{2-t} (t + d_0^2(M)t + d_0^2(M))}$$

for all $(x,t) \in M \times [0,2)$.

This sharpens the result of [BCP] where the coefficient in front of $|\nabla u|^2/u^2$ is not 1 and Ricci curvature is assumed to be bounded. Instead, even some singularity of the Ricci curvature is allowed.

Some possible problems to work on are:

- Characterize noncompact manifolds satisfying the sharp Li-Yau bound?
- sharp Li-Yau bound for the forward heat equation under Ricci flow?
- sharp Li-Yau bound in large time?
- Hamilton's entropy bound Theorem 5 under integral Ricci condition?

6 Matrix log gradient bounds

As mentioned, R. Hamilton extended the Li-Yau estimate to the log matrix form. On a general compact n-manifold $\mathbf{M}$, he proved the following

Theorem 14 ([Ha93]) *For any positive solution u to (2.1), there exist cons-tants B and C depending only on the geometry of $\mathbf{M}$ (in particular the diameter, the volume, and the curvature and covariant derivative of the Ricci curvature) such that if $t^{n/2}u \leqslant B$ then*

$$\nabla_i\nabla_j \ln u + \frac{1}{2t}g_{ij} + C\left(1 + \ln\left(\frac{B}{t^{n/2}u}\right)\right)g_{ij} \geqslant 0. \tag{6.1}$$

Moreover $C = 0$ if $\mathbf{M}$ has nonnegative sectional curvature and parallel Ricci curvature and the inequality holds for all positive solution u.

Joint with Xiaolong Li, we found an improved version of the above theorem by Hamilton.

Theorem 15 ([LZ]) *Let $(\mathbf{M}, g)$ be a closed Riemannian n-manifold and let $u : \mathbf{M} \times [0, T] \to \mathbf{R}$ be a positive solution to the heat equation (2.1). Suppose that the sectional curvatures of M are bounded by K and $|\nabla Ric| \leqslant L$, for some $K, L > 0$. Then*

$$\nabla_i\nabla_j \log u + \left(\frac{1}{2t} + (2n-1)K + \frac{\sqrt{3}}{2}L^{\frac{2}{3}} + \frac{1}{2t}\gamma(t, n, K, L)\right)g_{ij} \geqslant 0 \tag{6.2}$$

for all $(x, t) \in \mathbf{M} \times (0, T)$, where $\gamma(t, n, K, L)$ is

$$\sqrt{nKt(2 + (n-1)Kt)} + \sqrt{C_3(K + L^{\frac{2}{3}})t(1 + Kt)(1 + K + Kt)}$$

$$+ \left(2K(2 + (n-1)Kt) + \frac{3}{2}L^{\frac{2}{3}}(1 + (n-1)Kt)\right)diam,$$

$C_3 > 0$ depends only on the dimension n, and diam denotes the diameter of $(\mathbf{M}, g)$.

Notice that Hamilton's original inequality (6.1) has the term $C \log \left(\dfrac{B}{t^{\frac{n}{2}} u} \right)$, where B and C depend on the geometry of the manifold and B is greater than $t^{\frac{n}{2}} u$, which itself is an additional assumption. The constant C is equal to zero only when $\mathbf{M}$ has nonnegative sectional curvature and parallel Ricci curvature. Otherwise, for this log term, we do not have any definite control on the order q of $-t^{-q}$ coming out of this term, for general positive solutions, making this lower bound less practical. In Theorem 15, we manage to replace this term with a C/t term, with C depending only on K, L, and $diam$, which is of the correct order for t.

We also establish a similar result for a general compact Ricci flow.

Theorem 16 ([LZ]) *Let* $(\mathbf{M}, g(t))$, $t \in [0, T]$, *be a compact solution to the Ricci flow. Let* $u : \mathbf{M} \times [0, T] \to \mathbf{R}$ *be a positive solution to the heat equation* (3.2). *Suppose the sectional curvatures of* $(\mathbf{M}, g(t))$ *are bounded by* K *for some* $K > 0$. *Then*

$$\nabla_i \nabla_j \ln u + \left(\frac{1}{2t} + \frac{1}{t} \beta(t, n, K) \right) g_{ij} \geqslant 0, \tag{6.3}$$

where

$$\beta(t, n, K) = 4\sqrt{nKt} + C_2(K + 1)t + C_1 \sqrt{K} diam.$$

Here $C_1 > 0$ *is a numerical constant,* $C_2 > 0$ *depends only on the dimension and the non-collapsing constant* $v_0 = \inf\{|B(x, 1, g(0))|_{g(0)} : x \in M\}$, *and*

$$diam := \sup_{t \in [0, T]} diam(M, g(t)).$$

Similar upper bounds for $Hess \ln u$ can be found in [HZ]. The idea in that paper of using local extension of eigenvectors into vector fields is also instrumental in the proof of the preceding two theorems. This method allows us to avoid Hamilton's global tensor maximum principle.

Theorem 17 ([HZ]) *Let* $\mathbf{M}$ *be a Riemannian* n-*manifold with a metric* g.

(a) *Suppose* u *is a solution of*

$$\partial_t u - \Delta u = 0 \quad in \ \mathbf{M} \times (0, T].$$

Assume $0 < u \leqslant A$. *Then,*

$$t(\nabla_i \nabla_j u) \leqslant u(5 + Bt) \left(1 + \log \frac{A}{u} \right) \quad in \ \mathbf{M} \times (0, T],$$

where B is a nonnegative constant depending only on the L^∞-norms of curvature tensors, n and the gradient of Ricci curvatures.

(b) *Suppose u is a solution of*

$$\partial_t u - \Delta u = 0 \quad \text{in } Q_{R,T}(x_0, t_0) = B(x_0, R) \times [t_0 - T, t_0].$$

Assume $0 < u \leqslant A$. Then,

$$(\nabla_i \nabla_j u) \leqslant Cu \left(\frac{1}{T} + \frac{1}{R^2} + B \right) \left(1 + \log \frac{A}{u} \right)^2 \quad \text{in } Q_{\frac{R}{2}, \frac{T}{2}}(x_0, t_0).$$

where C is a universal constant and B is a nonnegative constant depending only on the L^∞-norms of curvature tensors, n and the gradient of Ricci curvatures in $Q_{R,T}(x_0, t_0)$.

Notice that $\nabla_i \nabla_j \ln u = \nabla_i \nabla_j u / u - \nabla_i u \nabla_j u / u^2$. So the above theorem together with Theorem 5 and Theorem 6 yield the respective upper bound for $\nabla_i \nabla_j \ln u$.

Hamilton's Theorem 14 is sharp when the sectional curvature is nonnegative and the Ricci curvature is parallel. The equality is reached by the heat kernel in the Euclidean space. It is noticed by Poon [Po] that this sharp matrix estimate is connected with the unique continuation property for the heat equation on such manifolds.

In the paper [LZ], we extended Hamilton's sharp matrix estimate (Theorem 14), $C = 0$ case for static metrics to the Ricci flow case. Our estimate does not require the parallel Ricci curvature condition and thus could be more applicable.

Theorem 18   *Let $(\mathbf{M}^n, g(t))$, $t \in [0, T]$, be a complete solution to the Ricci flow. Let $u : \mathbf{M}^n \times [0, T] \to \mathbf{R}$ be a positive solution to the heat equation*

$$u_t - \Delta_{g(t)} u = 0. \tag{6.4}$$

Suppose that $(\mathbf{M}^n, g(t))$ has nonnegative sectional curvature and $Ric \leqslant \kappa g$ for some constant $\kappa > 0$. Then

$$\nabla_i \nabla_j \log u + \frac{\kappa}{1 - e^{-2\kappa t}} g_{ij} \geqslant 0, \tag{6.5}$$

for all $(x, t) \in M \times (0, T)$.

In the same paper, this theorem is used to prove the unique continuation property (UCP) of the conjugate heat equation on those manifolds using monotonicity of a modified parabolic frequency function with the heat kernel of the forward heat equation as weight. Let us recall that a solution u satisfies (UCP) if vanishing at infinity order at one space time point implies vanishing everywhere. This result is in part inspired by the results in the papers [Lin90], [Po], [CM] and [BK]. See also [LW19] and [LLX22]. It is well known that (UCP) requires very sharp parameters in the monotonicity formula, which is provided by Theorem 18. An interesting question is whether the theorem still holds without the upper bound on the Ricci curvature. Also the parameters in the matrix Harnack estimates are likely not sharp except we have nonnegative sectional curvature and parallel Ricci curvature in the static case. It is desirable to find sharp ones.

Acknowledgments The author gratefully acknowledges the support of Simons' Foundation grant 710364 and Prof. Xiaolong Li for helpful suggestions. He also wishes to thank Professors Gang Tian, Jiayu Li, Zhenlei Zhang and Xiaohua Zhu for the invitation to write the paper and to the annual geometric analysis conference.

References

[AbGr]   Abresch U, Gromoll D, *On complete manifolds with nonnegative Ricci curvature*, J. Amer. Math. Soc. 3 (1990), no. 2, 355-374.

[An]   Anderson M, *On the topology of complete manifolds of nonnegative Ricci curvature*, Topology 29 (1990), no. 1, 41-55.

[ArBe]   Aronson D G, Bénilan P, *Régularité des solutions de l'équation des milieux poreux dans RN*, C. R. Acad. Sci. Paris Sér. A-B 288 (1979), no. 2, A103-A105.

[Ber]   Bernstein S N, *Sur la généralisation du probléme de Dirichlet. 1st, 2nd,* Math. Ann. 62 (1906), 253-271; Math. Ann. 69 (1910), 82-136.

[BCP]   Bailesteanu M, Cao X D, Pulemotov A, Gradient estimates for the heat equation under the Ricci flow. J. Funct. Anal. 258 (2010), no. 10, 3517-3542.

[BBG]   Bakry D, Bolley F, Gentil I, *The Li-Yau inequality and applications under a curvature-dimension condition*, arXiv:1412.5165.

[BL]   Bakry D and Ledoux M, *A logarithmic Sobolev form of the Li-Yau parabolic inequality*, Rev. Mat. Iber. 22(2006), no.2, 683-702.

[BNS]   Brue E, Naber A, Semola D, *Fundamental groups and the Milnor conjecture*, arXiv:2303.15347.

[BZ]   Bamler R, Zhang Q S, *Heat kernel and curvature bounds in Ricci flows with bounded scalar curvature*, Advan. Math. 319(2017), 396-450.

[Bre]   Brendle S, *A generalization of Hamilton's differential Harnack inequality for the Ricci flow*, J. Diff. Geom. 82 (2009), no. 1, 207-227.

[BK]   Baldauf J and Kim D, *Parabolic frequency on Ricci flows*, Int. Math. Res. Not. IMRN, to appear, arXiv:2201.05505v2.

[Cao]   Cao H D, *On Harnack's inequalities for the Kähler-Ricci flow*, Invent. Math. 109 (1992), no. 2, 247-263.

[CaNi]   Cao H D, Ni L, *Matrix Li-Yau-Hamilton estimates for the heat equation on Kähler manifolds*, Math. Ann. 331 (2005), no. 4, 795-807.

[Car]   Carron G, *Geometric inequalities for manifolds with Ricci curvature in the Kato class*, Ann. Inst. Fourier (Grenoble) 69 (2019), no. 7, 3095-3167.

[CTZ]   Cao H D, Tian G, Zhu X H, *Kähler-Ricci solitons on compact complex manifolds with $C_1(M) > 0$*, Geom. Funct. Anal. 15 (2005), no. 3, 697-719.

[Cx]   Cao X D, *Differential Harnack estimates for backward heat equations with potentials under the Ricci flow*, J. Funct. Anal. 255 (2008), no. 4, 1024-1038.

[CH]   Cao X D, Hamilton R S, *Differential Harnack estimates for time-dependent heat equations with potentials*, Geom. Funct. Anal. 19 (2009), no. 4, 989-1000.

[CZ]   Cao X D, Zhang Q S, *The Conjugate Heat Equation and Ancient Solutions of the Ricci Flow*, Adv. Math., 228 (2011), no. 5, 2891-2919.

[Chow]   Chow B, *The Ricci flow on the 2-sphere*, J. Diff. Geom. 33 (1991), no. 2, 325-334.

[ChH]   Chow B, Hamilton R S, *Constrained and linear Harnack inequalities for parabolic equations*, Invent. Math. 129 (1997), no. 2, 213-238.

[CW13]   Chen X X, Wang B, *On the conditions to extend Ricci flow(III)*, Int. Math. Res. Not. IMRN(2013), no. 10, 2349-2367.

[Chowetc]   Chow B, Lu P, Ni L, *Hamilton's Ricci flow*. American Mathematical Society, Providence, RI; Science Press, New York, 2006. xxxvi+608 pp.

[CoNa]   Colding T H, Naber A, *Sharp Hölder continuity of tangent cones for spaces with a lower Ricci curvature bound and applications*, Ann. of Math. 176 (2012), no. 2, 1173-1229.

[Cr]   Croke C, *Some isoperimetric inequalities and eigenvalue estimates*, Ann. Sci. Ecole Norm. Sup. 13 (1980), 419-435.

[CM]   Tobias H C and William P M, II. *Parabolic frequency on manifolds*, Int. Math. Res. Not. IMRN, 15(2022), 11878-11890.

[CY]   Cheng S Y and Yau S T, *Differential equations on Riemannian manifolds and their geometric applications*, Comm. Pure Appl. Math. 28(1975), no.3, 333-354.

[Dav]   Davies E B, *Heat Kernels and Spectral Theory*, Cambridge: Cambridge University Press, 1989.

[DWZ]   Dai X Z, Wei G F, Zhang Z L, *Local Sobolev constant estimate for integral Ricci curvature bounds*, Advan. Math. 325 (2018), 1-33.

[Gal]   Gallot S, *Isoperimetric inequalities based on integral norms of Ricci curvature*, Asterisque, 157-158 (1988), 191-216.

[GM]   Garofalo N and Mondino A, *Li-Yau and Harnack type inequalities in $RCD^*(K; N)$ metric measure spaces.* Nonlinear Anal. 95(2014): 721-734.

[Gri]   Grigor'yan A A, *The heat equation on noncompact Riemannian manifolds*, Mat. Sb., 182 (1991), no.1, 55-87; Translation in Math. USSR-Sb., 72(1992), no.1, 47-77.

[Gro]   Gromov M, *Paul Levy's Isoperimetric Inequality*, Appendix C in Metric Structures for Riemannian and Non-Riemannian Spaces, Birkhauser, 2001.

[Ha93]   Hamilton R S, *A matrix Harnack estimate for the heat equation*, Comm. Anal. Geom. 1 (1993), no. 1, 113-126.

[Ha4]   Hamilton R S, *The Ricci flow on surfaces*, in: Isenberg J A, Mathematics and General Relativity, Amer. Math. Soc., Providence, RI, 1988, 237-262.

[Ha5]   Hamilton R S, *The Harnack estimate for the Ricci flow*, J. Differential Geom. 37 (1993), no. 1, 225-243.

[HY]   Hua B B, Yang W H, *Liouville theorems for ancient solutions of subexponential growth to the heat equation on graphs*, arXiv:2309.17250.

[HZ]   Han Q, Zhang Q S, *An upper bound for Hessian matrices of positive solutions of heat equations*, J. Geome. Anal. 26 (2016), 715-749.

[KuZh]   Kuang S L, Zhang Q S, *A gradient estimate for all positive solutions of the conjugate heat equation under Ricci flow*, J. Funct. Anal. 255 (2008), no. 4, 1008-1023.

[Li]   Li P, *Large time behavior of the heat equation on complete manifolds with nonnegative Ricci curvature*, Ann. Math. 124 (1986), no. 1, 1-21.

[Lib]   Li P, *Geometric Analysis*. Cambridge Studies in Advanced Mathematics, 134. Cambridge University Press, Cambridge, 2012, x+406 pp.

[Liu]   Liu G, *3-manifolds with nonnegative Ricci curvature.* Invent. Math. 193 (2013), no. 2, 367-375.

[LLX22]   Li C H, Li Y, and Xu K R, *Parabolic frequency monotonicity on Ricci flow and Ricci-harmonic flow with bounded curvatures.* J. Geom. Anal., to appear, arXiv:2205.07702.

[LW19] Li X L and Wang K, *Parabolic frequency monotonicity on compact manifolds*, Calc. Var. Partial Diff. Equa. 58(2019), no.6, Paper No. 189, 18.

[LX]   Li J and Xu X, *Differential Harnack inequalities on Riemannian manifolds I: linear heat equation*. Adv. Math. 226(2011), no.5, 4456-4491.

[LZ]   Li X L, Zhang Q S, *Matrix Li-Yau-Hamilton estimates under Ricci Flow and parabolic frequency*, arXiv:2306.10143.

[Lij]   Li J Y, *Gradient estimate for the heat kernel of a complete Riemannian manifold and its applications*, J. Funct. Anal. 97(1991), 293-310.

[Lin90] Lin F H, *A uniqueness theorem for parabolic equations*. Comm. Pure Appl. Math. 43(1):127-136, (1990).

[LinZ] Lin F H and Zhang Q S, *On ancient solutions of the heat equation*. Comm. Pure Appl. Math. 72(2019), no.9, 2006-2028.

[LWH+] Lei Z, Wang Z Q, Hua B B, et al., *Lectures on mathematical and physical equations*. Fudan Lecture Notes, in Chinese, unpublished.

[LT]   Li P, Tian G, *On the heat kernel of the Bergmann metric on algebraic varieties*, J. Amer. Math. Soc., 8 (1995), 857-877.

[LY]   Li P, Yau S T, *On the parabolic kernel of the Schrödinger operator*, Acta Math. 156 (1986), no. 3-4, 153-201.

[Ma]   Mabuchi T, *Heat kernel estimates and the Green functions on Multiplier hermitian manifolds*, Tohoku Math. J. 54 (2002), 259-275.

[Mo]   Modica L, *A gradient bound and a Liouville theorem for nonlinear Poisson equations*, Comm. Pure Appl. Math. 38 (1985), no. 5, 679-684.

[Mos]   Mosconi S, *Liouville theorems for ancient caloric functions via optimal growth conditions*, Proc. Amer. Math. Soc. 149(2021), no.2, 897-906.

[Ni]   Ni L, *A note on Perelman's LYH-type inequality*, Comm. Anal. Geom. 14 (2006), no. 5, 883-905.

[Pa]   Pan J Y, *A proof of Milnor conjecture in dimension 3*, J. Reine Angew. Math. 758 (2020), 253-260.

[P1]   Perelman G, *The entropy formula for the Ricci flow and its geometric applications*, http://arxiv.org/abs/math/0211159 (2002).

[Po]   Poon C C, *Unique continuation for parabolic equations*, Comm. Partial Differential Equations, 21(1996), no. 3-4, 521-539.

[PW1]   Petersen P, Wei G F, *Relative volume comparison with integral curvature bounds*, Geom. Funct. Anal. 7 (1997), no. 6, 1031-1045.

[PW2]   Petersen P, Wei G F, *Analysis and geometry on manifolds with integral Ricci curvature bounds. II*, Trans. Amer. Math. Soc. 353 (2001), no. 2, 457-478.

[QZZ] Qian Z, Zhang H C and Zhu X P, *Sharp spectral gap and Li-Yau's estimate on Alexandrov spaces*, Math. Z., 273(2013), no. 3-4, 1175-1195.

[Ro] Rose C, *Li-Yau gradient estimate for compact manifolds with negative part of Ricci curvature in the Kato class*, Ann. Global Anal. Geom. 55 (2019), no. 3, 443-449.

[Sa] Saloff-Coste L, *A note on Poincaré, Sobolev, and Harnack inequalities*, Inter. Math. Res. Notices, 1992(1992), no.2, 27-38.

[So] Sormani C, *Nonnegative Ricci curvature, small linear diameter growth and finite generation of fundamental groups*, J. Differential Geom. 54 (2000), no. 3, 547-559.

[SW] Song X Y, Wu L, *Li-Yau gradient estimates on closed manifolds under bakry-emery ricci curvature conditions*, arXiv:2204.12851.

[SWZ] Song X Y, Wu L, Zhu M, *A direct approach to sharp Li-Yau Estimates on closed manifolds with negative Ricci lower bound*, arXiv:2307.03879.

[SZ] Souplet P, Zhang Q S, *Sharp gradient estimate and Yau's Liouville theorem for the heat equation on noncompact manifolds*, Bull. London Math. Soc. 38 (2006), no. 6, 1045-1053.

[TZq1] Tian G, Zhang Q S, *Isoperimetric inequality under Kähler Ricci flow*, Amer. J. Math. 136 (2014), no. 5, 1155-1173.

[TZq2] Tian G, Zhang Q S, *A compactness result for Fano manifolds and Kähler Ricci flows*, Math. Ann. 362 (2015), no. 3-4, 965-999.

[TZz] Tian G, Zhang Z L, *Regularity of Kähler Ricci flows on Fano manifolds*, Acta Math. 216(2016), no.1, 127-176 (2016).

[Wan] Wang F Y, *Gradient and Harnack inequalities on noncompact manifolds with boundary*, Pacific J. Math., 245(2010), no.1, 185-200.

[WanJ] Wang J P, *Global heat kernel estimates*, Pacific J. Math. 178 (1997), no. 2, 377-398.

[Ya] Yau S T, *Harmonic functions on complete Riemannian manifolds*, Comm. Pure Appl. Math. 28 (1975), 201-228.

[Yan] Yang D, *Convergence of Riemannian manifolds with integral bounds on curvature. I*, Ann. Sci. Ec. Norm. Super. 25 (1992), 77-105.

[YZ] Yu C J, Zhao F F, *Li-Yau multiplier set and optimal Li-Yau gradient estimate on hyperbolic spaces*, arXiv:1807.05709.

[Z06] Zhang Q S, *Some gradient estimates for the heat equation on domains and for an equation by Perelman*, Int. Math. Res. Not., 39 pages Art. ID 92314, 39, 2006.

[Z12] Zhang Q S, *Bounds on volume growth of geodesic balls under Ricci flow*, Math. Res. Letters, 19(2012), no.1 (2012), 245-253.

[Z21]   Zhang Q S, *A sharp Li-Yau gradient bound on compact manifolds*, arXiv:2110.08933 v2, 2021.

[ZZ1]   Zhang Q S and Zhu M, *Li-Yau gradient bounds on compact manifolds under nearly optimal curvature conditions*, J. Func. Anal. 275 (2018), no.2, 478-515.

[ZZ2]   Zhang Q S and Zhu M, *Li-Yau gradient bound for collapsing manifolds under integral curvature condition*, Proc. Amer. Math. Soc. 145(2017), 3117-3126, 2017.